The Future of Mobility

Scenarios for the United States in 2030
Johanna Zmud, Liisa Ecola, Peter Phleps, Irene Feige

The research described in this report was sponsored by the Institute for Mobility Research (ifmo) and conducted in the Transportation, Space, and Technology Program within RAND Justice, Infrastructure, and Environment.

As a part of the agreement under which the research described in this document was funded, RAND grants the Institute for Mobility Research a nonexclusive, royalty-free license to duplicate and distribute this publication in any medium, either commercially or non-commercially.

Library of Congress Cataloging-in-Publication Data is available for this publication.
ISBN: 978-0-8330-8139-1

The RAND Corporation is a research organization that develops solutions to public policy challenges to make communities throughout the world safer and more secure, healthier and more prosperous.

RAND's publications do not necessarily reflect the opinions of its research clients and sponsors.

Support RAND—make a tax-deductible charitable contribution at www.rand.org/giving/contribute.html.

RAND OFFICES
SANTA MONICA, CA • WASHINGTON, DC • PITTSBURGH, PA • NEW ORLEANS, LA • JACKSON, MS • BOSTON, MA • DOHA, QA • CAMBRIDGE, UK • BRUSSELS, BE

www.rand.org

Foreword

Mobility: Why does it matter? People seldom travel just for the sake of it—they do so for the purpose of work or leisure, and for a host of other reasons. The world over, mobility is associated with increasing economic output, higher standards of living, and personal freedom—and the diversity of lifestyles that such freedom entails. While it is true that emerging technology can, in some cases, afford virtual opportunities that substitute for mobility—as, for example, when electronic communication replaces face-to-face interaction—there is no doubt that mobility will continue to play a major role in societies of the future. A clearer understanding of how mobility is likely to change helps policymakers, businesses, and individuals to make better-informed decisions about which kinds of transport services to use, which technologies and equipment to choose, and what infrastructure to invest in.

The Institute for Mobility Research (ifmo) provides an invaluable service to decisionmakers in the sphere of transport and transport investment by publishing scenarios in a series titled "The Future of Mobility." The scenarios for Germany appear on a regular five-year basis and have earned a reputation for being a source of sound and valuable information. In the modern age, local contexts—and even national ones—are proving increasingly inadequate for formulating an understanding of transport. Thus the decision by the institute to carry out a similar scenario study for the United States, for the year 2030, was a logical next step. The opportunity that international comparisons provide for assisting decisionmaking adds even more value to the scenarios. Although transport systems differ from each other in important respects, the scenarios need a common analytical core. The present study strikes a fine balance between accounting for distinctive national characteristics and acknowledging shared conceptual underpinnings with the established German scenarios.

Dr. Andreas Kopp
Lead economist, The World Bank Group
Member, ifmo board of trustees

Preface

The future of mobility in the United States is important to policy- and decisionmakers. Without some ideas about how and how much Americans will travel in the future, it is difficult to know whether the U.S. roadway infrastructure will be adequate, whether existing funding sources will be sufficient, and how new circumstances will change mode shares (percentages of travelers using different types, or modes, of transportation). Instead of trying to predict these situations, or extrapolate from existing trends, the research team on the project reported here used a scenario approach to develop two distinct alternative futures for the country. Data were based on expert opinions about the long-term future in five areas: demographics, economics, energy, transportation funding, and technology.

The research reported here was sponsored by the Institute for Mobility Research, known by its German abbreviation ifmo. ifmo has conducted several scenario exercises for Germany and engaged the RAND Corporation to execute a scenario study for the United States. The results should be of interest to policy- and decisionmakers concerned with the long-term future of transportation in the United States. For the Transportation Research Board, RAND is conducting other long-term strategic studies, looking at options for adopting alternatively fueled vehicles, incorporating new technologies into the transportation system, and evaluating the impact that sociodemographic changes can have on travel demand.

The RAND Transportation, Space, and Technology Program

The research reported here was conducted in the RAND Transportation, Space, and Technology Program, which addresses topics relating to transportation systems, space exploration, information and telecommunication technologies, nano- and biotechnologies, and other aspects of science and technology policy. Program research is supported by government agencies, foundations, and the private sector. This program is part of RAND Justice, Infrastructure, and Environment, a division of the RAND Corporation dedicated to improving policy and decision-making in a wide range of policy domains, including civil and criminal justice, infrastructure protection and homeland security, transportation and energy policy, and environmental and natural resource policy.

Questions or comments about this report should be sent to the project leader, Johanna Zmud (Johanna_Zmud@rand.org). For more information about the Transportation, Space, and Technology Program, see http://www.rand.org/transportation or contact the director at tst@rand.org.

The Institute for Mobility Research

The Institute for Mobility Research is a research facility of the BMW Group. It deals with future developments and challenges relating to mobility across all modes of transport, with automobility being only one aspect among many. Taking on an international perspective, ifmo's activities focus on social science and sociopolitical, economic, and ecological issues, and also extend to cultural questions related to the key challenges facing the future of mobility. The work of the institute is supported by an interdisciplinary board of renowned scientists and scholars and by representatives of BMW, Deutsche Bahn, Lufthansa, MAN, Siemens, and the World Bank.

Contents

Summary

Research Question

What might we expect for the future of mobility in the United States in 2030? Responses to this question will help transportation agencies at federal, state, and local levels to better prepare for the future. Although there is a legacy planning process to guide transportation decisions, long-range transportation planning involves many difficult choices, especially in an era of constrained resources. Which modes of transportation should be prioritized? Which investments should be funded? Which are the most-important trends to monitor over time? How are demographics, economics, and travel behavior likely to interact over time? These questions are hard to answer, particularly because transportation planners and policymakers must make decisions within a time horizon that extends 30 to 50 years into the future.

Although we know that the nation's mobility (how people are capable of traveling from point to point) will be considerably different in 2030, figuring out how it will be different is a significant challenge. Substantial change tends to happen relatively slowly, but a long-term future can look very different from the situation today. For example, the U.S. Census Bureau projects that the U.S. population will grow from 308 million in 2010 to about 360 million in 2030. This is a big change from today's number. Total miles traveled will also increase substantially as the population increases. However, the projected population growth will happen more slowly than the growth recorded in the past 50 years. How slowly the population grows will depend on the interactions of three under-lying determinants: fertility, mortality, and net immigration. So what is important for long-term future policy- and decisionmaking regarding mobility is *not* how much the population will grow but *where* it will grow and *who* makes up that growth, as well as *what* will be the summation of the billions of their individual decisions about their mobility needs and wants. Demographic changes are only one of many different areas influencing future mobility.

Answers to our research question cannot be reliably addressed through straight-line trend analysis or improved travel demand forecast models. These approaches are lacking because the data and information to support long-term thinking about the future of mobility are uncertain, incomplete, evolving, or conflicting. Instead, we have applied scenario techniques, which are increasingly being used to deal with opportunities and risks of complex long-term issues. As we look ahead to 2030, multiple mobility futures are possible. Policy leaders face a big challenge in keeping people and goods moving today while reducing or avoiding negative consequences for the future. The relationship between today's situation and a long-term future outcome is not linear. It is not even relevant to study the two points in time—now and then. It takes a systematic process of identifying possible, plausible futures and then of understanding the paths leading to those alternative futures.

Our study, which was a collaboration of RAND and the Institute for Mobility Research (ifmo), focused on long-term scenarios for passenger travel, which includes travel by car, transit, domestic air, and intercity rail. Long-term scenarios in this area are multilayered and complex, being influenced by demographics, economics, energy, transportation funding and supply, and technology. How these forces play out over the next 20 years will depend on whether and how policymakers and other decisionmakers sort out and address current and upcoming challenges. Although we cannot know these outcomes in advance, we can apply scenario planning to develop plausible mobility futures that can be used to anticipate and prepare for change.

Methodology

To develop alternative scenarios of the future of mobility, we applied a process that combined expert opinion gathered in workshops, impact analysis, consistency analysis, and cluster analysis. The study began with identifying five influencing areas and descriptors (variables of interest) within each area. Then RAND and ifmo convened five workshops, one for each influencing area: demographics, economics, energy, transportation funding and supply, and technology.

Six to eight subject-matter experts from government, academia, nonprofit organizations, and consulting firms were involved in each workshop, for a total of 37 individuals who brought considerable substantive experience in a variety of fields and disciplines. At each workshop, experts were asked for projections for each descriptor for 2030, along with their assumptions regarding the projection and their qualitative estimate of its impact on mobility. Where there was little uncertainty and high consensus, only one projection per descriptor was identified. Otherwise, two or three alternative projections surfaced.

The descriptors and projections were subjected to a cross-impact analysis and consistency analysis to identify relationships between the descriptors; these were then input into a computer support system. Cluster analysis was then used to group them into distinct scenario frameworks. Two scenarios were produced, No Free Lunch and Fueled and Freewheeling. A panel of 27 outside experts validated the scenarios through an online Delphi system, ExpertLens. The resulting scenario narratives were developed based on the assumptions and projections that surfaced during the expert workshops.

The Scenarios

The scenarios provide two distinct perspectives on the future of mobility in the United States in 2030. Each future represents a particular trajectory to arrive at the outcome. One path recognizes that climate-change effects, if severe and observable by enough Americans, will shift public sentiment to heavily favor regulation to mitigate greenhouse-gas (GHG) emissions. This policy direction, in addition to other global influences, causes a very high price of oil. Americans are driving less and using alternatives to conventional vehicles. On a rather different trajectory, abundant energy and cheap oil because of new supplies, technology, and global demand drive down the price of oil. The economy is booming, and Americans are driving more.

In this section, we briefly explore the United States in 2030 based on the two scenarios. Each provides a different future driven by a particular series of developments over the next two decades. Figures S.1 and S.2 provide a visual snapshot of these two scenarios.

Scenario 1: No Free Lunch

In 2030, several factors will have combined to bring about strengthened regulations to reduce dependency on oil and GHG emissions. Oil prices, rising for years, will have hit an all-time high. Two decades of undeniable evidence of climate change will have sparked changing attitudes among the public and business community to effect legislative change. National GHG-reduction policies will have been implemented. This legislation will have spurred innovation in the energy domain and the uptake of renewable and alternative fuels. Gross domestic product (GDP) and oil consumption will no longer be coupled; oil consumption will be down, and the supply will be constrained. New zoning restrictions will have created greater densities in urban and suburban areas, which, in turn, will have increased public transit use. The "young elderly" will continue to drive but favor alternatively fueled vehicles (AFVs) for cost savings and vehicles equipped with advanced driver-assistance systems for safety. Road pricing will be prevalent as a source of needed revenues to maintain and expand the surface transportation system and as a disincentive to use the system, which will have caused tangible reductions in congestion. The United States will address the effects of climate change with regula-tion, having reached a national consensus on its causes and effects, and this will have had a positive rather than a negative effect on the economy.

Scenario 2: Fueled and Freewheeling

In 2030, lower-than-anticipated oil prices in the preceding 15 years will have led to a future quite similar to life in the 1980s and 1990s. The economy will be booming, and the anticipated harms from climate change will not have unduly affected Americans. Oil prices will have stabilized as the economy pulls out of the Great Recession caused by the financial crisis of 2008. Lacking external pressures, elected officials will not have pursued changes in U.S. energy policy. The share of AFVs in the fleet will be low, and inexpensive fuel will also have made driving relatively inexpensive. This situation will have induced more suburbanization. The market for suburban living will be thriving based on demand from young families with money to spend on freestanding houses and new cars. Climate change will be happening, but the effects will be localized. Policies to mitigate climate change will have been adopted only by highly affected states or cities with the most-committed populaces. A reluctance to raise taxes will have left the transportation sector severely underfunded. A few states will have adopted modified mileage-fee systems, and a few big cities will have imposed congestion charges. But overall, the condition of U.S. roads and bridges will be getting worse, although the extent of the decline will vary greatly. Technology will substitute for some travel but not enough, and congestion will have grown.

The price of oil is a major driver in both scenarios, along with the level of environmental regulation and the amount of highway revenues and expenditures.

Acknowledging that scenarios can be constrained by what is plausible, believable, or imaginable today, we crafted two wild-card scenarios to provoke "thinking about the unthinkable." These assume that certain events have broken with otherwise-foreseeable trends.

One wild card is based on the possibility that China experiences a major debt crisis and ensuing economic stagnation, with economic and demographic impacts that profoundly affect the United States. The other assumes that autonomous vehicles, currently unavailable commercially and thought by our experts to be several decades away, experience cost reductions that make them marketable much sooner than expected, with attendant effects on transportation.

$ 90

2

Fueled and Freewheeling

When cheap and abundant energy, relatively low oil prices, and a lack of regulation combine to create high transport demand

High per capita VMT • Significant congestion • High immigration • Low unemployment • Cheap to drive • More cars • Crumbling infrastructure • Demand for air travel • New home sales • Fuel-efficient cars • Suburbanization • Geographic winners and losers

Implications

For Future Mobility

For each scenario, we developed estimates of passenger-miles traveled (PMT) in 2030 for four transportation modes: vehicle, transit, domestic air, and intercity rail. Our analysis takes into account the descriptors' influence on travel demand and the strength of that influence. Under both scenarios, the number of PMT in the United States has grown by 2030. Much of the increase is due to population growth. However, the increase is greater in Fueled and Freewheeling than in No Free Lunch by almost a factor of four, indicating the influence of factors other than demographics. PMT increased by 22 percent between 2010 and 2030 in Fueled and Freewheeling and by 6 percent in No Free Lunch (while population increased by 17.3 percent). On a per capita basis, though, PMT actually declined in the latter scenario by 9.5 percent even as total PMT increased. This is due to decreases in daily travel among certain population groups, such as young adults, older persons, and the technology-connected.

In both scenarios, growth in travel by air (68-percent increase in Fueled and Freewheeling, 37 percent in No Free Lunch) dwarfs the growth for highway miles of travel (16 percent and 2 percent, respectively). In the No Free Lunch scenario, economic growth has pushed air transport up, but high oil prices, in combination with additional carbon dioxide (CO_2) emission trading costs, have increased ticket prices.[1] So air travel demand is diminished from what it would be in the Fueled and Freewheeling scenario. In the Fueled and Freewheeling scenario, the significant growth is due to stronger economic growth and low oil prices. These two influencing areas, along with operational efficiencies, have caused airfares to grow more slowly than inflation, which drives substantial demand. As might be expected, growth rates for transit are robust under No Free Lunch (30 percent total and 11 percent per capita) and modest under Fueled and Freewheeling (17 percent total and –1 percent per capita). In 2030, in both scenarios, intercity rail remains a negligible contributor to total PMT.

For Transportation Agencies

Our two scenarios describe different mobility futures. The scenarios are descriptive, not normative—neither is put forward as the ideal path for the future of mobility in the United States. In addition, our study did not address the likelihood of one particular outcome versus another. The scenarios are instead indicative of a range of "plausibilities." By making potential long-term consequences more vivid, scenarios can support public policy by helping planners and policymakers at different levels of government envision what the future might bring.

---

[1] Emission trading costs are costs associated with actions and strategies used to reduce emissions in order to stay below a government-set limit (the cap) under a cap-and-trade policy scheme.

Our analysis revealed three driving forces as being significant in this regard: (1) the price of oil, (2) the development of environmental regulation, and (3) the amount of highway revenues and expenditures. The price of oil is exogenous; transportation policymakers have virtually no leverage over it. The other two drivers are well within the purview of transportation policy at all levels of government. In applying the scenarios in agencies' planning activities, we identified three possible approaches: (1) identifying early warning signs; (2) determining opportunities, risks, and contingencies; and (3) reviewing strategic options against the scenarios.

Conclusions

From our research, we find the following:

The future of mobility in the United States in 2030 is uncertain. This project created two scenarios, No Free Lunch and Fueled and Freewheeling, to illustrate the paths that may result from interconnected effects of market, policy, and consumer forces.

No Free Lunch describes a future in which the United States has strengthened regulations to reduce dependency on oil and GHG emissions, which results in greater investment in research and development (R&D) on AFVs, increased public transit ridership, greater reliance on road pricing, and lower levels of car ownership. Fueled and Freewheeling describes a future in which the economy is booming and a reluctance to raise taxes is prevalent, which results in high car ownership and steadily increasing congestion.

The study identified three critical uncertainties, or driving forces, that cause one path to emerge over another: the price of oil, the development of environmental regulation, and the amount of highway revenues and expenditures. Of these, the most critical is oil price.

The potential for transportation policymakers and other decisionmakers to influence the price of oil is limited. However, they will have greater opportunity to leverage the other key drivers.

Acknowledgments

We thank the many experts outside RAND who contributed their time at the expert workshops. They were gracious with their time, as well as their honest opinions about future projections in their respective areas of expertise. Our demographic experts are John Cromartie, Economic Research Service, U.S. Department of Agriculture; Ryan D. Edwards, Queens College, City University of New York; B. Lindsay Lowell, Georgetown University; Joyce Manchester, Congressional Budget Office; Nancy McGuckin, independent consultant in travel behavior; and Jeffrey S. Passel of the Pew Hispanic Center. In economics, our experts are Paul Bingham, CDM Smith; Greg Bischak, Community Development Financial Institutions Fund, U.S. Department of the Treasury; Luca Flabbi, Georgetown University; Andreas Kopp, World Bank; Marika Santoro, formerly of the Congressional Budget Office; Sita Nataraj Slavov, American Enterprise Institute; Michael Toman, World Bank; and John V. Wells, U.S. Department of Transportation. Our energy experts are Austin Brown, National Renewable Energy Laboratory, U.S. Department of Energy; Carmine Difiglio, Office of Policy and International Affairs, U.S. Department of Energy; Charles K. Ebinger, Brookings Institution; Jim Kliesch, Union of Concerned Scientists; Joshua Linn, Resources for the Future; Michael Shelby, Office of Transportation and Air Quality, U.S. Environmental Protection Agency; James T. Turnure, Energy Information Administration; and Jacob W. Ward, Office of Energy Efficiency and Renewable Energy, U.S. Department of Energy. In transportation funding and supply, our experts are Susan J. Binder, Cambridge Systematics; John W. Fischer, formerly of the Congressional Research Service; Emil H. Frankel, Bipartisan Policy Center; Art Guzzetti, American Public Transportation Association; Phillip Herr, U.S. Government Accountability Office; Valerie J. Karplus, Massachusetts Institute of Technology; Bruce Schaller, New York City Department of Transportation; and Mary Lynn Tischer, Federal Highway Administration. Finally, our technology experts are Steven H. Bayless, Intelligent Transportation Society of America; Matthew Dorfman, D'Artagnan Consulting; Frank Douma, University of Minnesota; Philip G. Gott, IHS Global Insight; Alain L. Kornhauser, Princeton University; Greg Krueger, Science Applications International Corporation; and Ted Trepanier, INRIX.

In addition, a group of experts provided their opinions about the scenarios via the ExpertLens tool; their anonymous participation precludes our thanking them publicly, but their input was extremely valuable.

We also thank the RAND researchers who developed background material for the expert workshops: Peter B. Brownell for demographics, Thomas Light for economics, Constantine Samaras and Paul Sorensen for energy, Nidhi Kalra and Jan Osburg for technology, and Paul Sorensen for funding. Their contributions helped set the stage for the expert projections and ensured that discussions began with a similar set of understanding of past trends. Keith Crane and David Yang provided a useful reality check on the China-based wild-card scenario. Other RAND staff helped greatly by taking detailed notes: Jeffrey Garnett at four workshops (demographics, economics, energy, and transportation funding) and Duncan Amos at the technology workshop.

Dmitry Khodyakov graciously walked us through how to set up an ExpertLens expert elicitation and did a superb job implementing one. He was assisted by Samuel Silver and Kiet Lieng. In addition, we thank Tobias Kuhnimhof of ifmo, who participated in several workshops and assisted with reviewing the final report.

We thank our two reviewers—Robert J. Lempert, senior scientist at RAND and director of the RAND Frederick S. Pardee Center for Longer Range Global Policy and the Future Human Condition, and Patricia L. Mokhtarian, professor of civil and environmental engineering and director of the Telecommunications and Travel Behavior Research Program at the Institute of Transportation Studies at the University of California, Davis—for many valuable suggestions on improving the draft report.

Last but not least, we thank some of RAND's very capable administrative assistants who made the travel and workshops run smoothly: Kate Piacente, Alexander Chinh, and Andria Tyner in the Arlington, Virginia, office and Donna Mead and Vera Juhasz in the Santa Monica office. They were also supported by RAND's very capable facility staff in both offices. Their able support allowed us to concentrate on the workshops, and we appreciate the effort to set up six workshops in a short period of time.

Chapter One
Introduction

For decades, transportation infrastructure in the United States has been built and maintained primarily to serve people and their cars. Starting just after World War II, the number of miles driven annually on America's roads steadily increased. The rising numbers were related to societal shifts, such as women joining the workforce, families moving to the suburbs, and the greater affordability of more cars for more people. Then, after the turn of the century, something changed: Americans began driving fewer miles—an unexpected development. Why this has occurred is, as of yet, not fully known. Reasons may include repercussions from the economic crisis of 2007-2008 and changing attitudes among young people about driving. This situation illustrates the fact that the future is unpredictable and shaped by many interacting factors.

This is where scenario planning comes into play. We use the term *scenario* to refer to a plausible combination of possible long-term future developments. Scenario planning is the development of one or more scenarios via a methodology that incorporates multiple possible future outcomes. The contribution of scenario planning is to help us consider a wider range of potential futures than those that would be predicted from the extrapolation of past trends or from a single set of projections.

The advantage to using scenarios in designing transportation policy is to foster discussion and analysis of possible outcomes that may not be obvious when using more-conventional tools, such as forecasting and modeling. Scenarios encourage transportation planners and policymakers to consider a wide range of possible, plausible futures and the paths leading to those futures. Decisions made in the short term can affect whether one scenario becomes more plausible than another, and scenarios can help identify "early warning signs" that can indicate which scenario has become more likely.

Study Objectives

This study applied scenario planning to answer this question: What might we expect for the future of mobility in the United States in 2030? We define mobility as the ability to travel from one location to another, regardless of mode or purpose. Instead of using trend analysis or quantitative forecasts to answer this question, we used scenario planning because it provides a structured method to explore the many ways in which mobility could evolve and then to examine what those possible alternative paths may imply about future mobility.

Our goal is not to predict the future—obviously, an impossible task—but to look at how various factors might affect mobility when combined in different ways. Our focus is largely on U.S. passenger travel (that is, personal daily travel via vehicle, transit, domestic air, and intercity rail). The goal is to better understand how a combination of factors can affect total mobility. For example, oil prices have a substantial effect on the amount of driving because drivers are sensitive to the price of gasoline. However, the choice to drive may also be influenced by other factors, such as investments in public transit systems or an economic downturn, that are taking place simultaneously.

To answer this question, RAND collaborated with the Institute for Mobility Research (ifmo) to apply a methodology that distills experts' projections in a variety of areas into scenarios that form plausible and consistent stories about the future. The use of scenarios to evaluate multiple potential futures is a technique first developed by RAND researchers in the 1960s (Kahn and Wiener, 1967) and has been considerably modified and expanded in the ensuing years. Börjeson et al. (2006) provide a simple typology of the many uses for which scenarios have been developed over the years. The technique we use here would be classified as an *external explorative* scenario. *External* means that it focuses on external factors, rather than what can happen if a particular actor takes a certain action. *Explorative* means that it seeks to understand what can happen in the future, rather than what will happen or how a certain target can be reached.

Many methodologies are available to develop scenarios, as discussed in Amer, Daim, and Jetter (2013). Cross-impact analysis is one means of developing the links between various factors, and consistency analysis is a means of ensuring that the many individual predictions that make up a scenario are internally consistent. Both of these tools can be used in qualitative, as well as quantitative, inputs.

The scenario methodology used in this project was developed by ifmo using a scenario framework presented in Gausemeier, Fink, and Schlake (1998). The online Risk Assessment and Horizon Scanning (RAHS) tool operationalizes the steps of the process shown in Figure 1.1. (More information about RAHS is contained in Appendix A.)

This study built on prior ifmo research that developed scenarios for Germany for 2020, 2025, and 2030. In that research, updates every five years have allowed the projections in each influencing area to be confirmed or adapted based on current contexts. Long-term planning horizons for transportation are typically several decades, for national and regional transportation planning, as well as for industry. For this report, we selected 2030 as the forecast year, in part for consistency with the earlier German work. Ideally, this report will become one of a series of reports that can be likewise updated or that use similar methodologies for other countries.

Creating the Scenarios

Our methodological approach represents a state-of-the-art scenario process while recognizing that scenarios may be developed using several different approaches. Our approach combined expert opinion, gathered via both in-person workshops and an online Delphi technique, with impact analysis, consistency analysis, and cluster analysis using specialized computer tool support. Even though it relies more on substantive expertise than on formal research and modeling, the approach was highly empirical. We describe it briefly in this section; Figure 1.1 summarizes the approach. More details are provided in Appendix A.

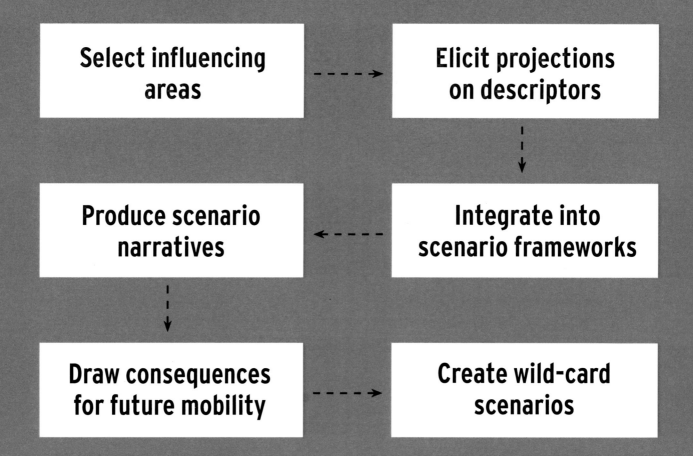

Figure 1.1. Six-Step Scenario Approach: From System Influences to Scenarios

Select Influencing Areas

The scenario process begins by defining three key study parameters: (1) topic (the future of mobility), (2) geographic scope (United States), and (3) time horizon (2030). The research team identified five influencing areas and specific descriptors to fit the study parameters. An influencing area is a broad topic area that is thought to affect mobility. The five in this study are demographics, economics, energy, transportation funding and supply, and technology. These were selected based on the German work, as well as concurrent RAND research for the Transportation Research Board.[1] For each area, we then identified descriptors, which are, in most cases, quantifiable metrics that represent one specific element within the influencing area. For example, among the descriptors identified for energy were the price of a barrel of oil and the percentage of new-vehicle purchases in 2030 that would be AFVs. A full list of the 32 descriptors can be found in Appendix A. The research team produced a paper on past and current trends for each influencing area; these papers are available in a separate volume of appendixes.

Elicit Projections on Descriptors

The research team held one workshop for each influencing area to gather expert opinion on projections for each descriptor. We defined a projection as an estimate of future possibilities based on past and current trends. We invited six to eight prominent outside experts (jointly identified by RAND and ifmo and listed in Appendix B) to attend one of the five workshops. Prior to each workshop, each expert received the paper on trends in his or her influencing area. At each workshop, using facilitated discussion, experts were asked for projections for each descriptor in 2030, clarifying that we were not asking them to extrapolate from past trends but rather to consider a variety of factors that they thought might influence the descriptor. For each descriptor, the experts provided between one and three projections. For example, we asked the energy experts to project the price of a barrel of oil in 2030, which led to estimates of high, medium, and low prices, all of which were considered plausible depending on circumstances. The number of projections depended on the degree of consensus on likely futures among the experts. We also asked the experts to provide their reasoning (or assumptions) for the projection. For example, those who projected a relatively low oil price in 2030 cited reduced global demand for oil, while those who projected a higher price mentioned political unrest in oil-producing regions and increased demand from developing countries.

---

[1] RAND researchers are leading three long-range projects for the National Cooperative Highway Research Program 20-83 series, looking at the impact of adopting alternatively fueled vehicles (AFVs), incorporating new technologies into the transportation system, and the impact of sociodemographic changes.

Integrate into Scenario Frameworks

The research team used two distinct types of analysis—cross-impact analysis and consistency analysis—to identify the values that would group descriptors and projections into distinct scenario frameworks (see Amer, Daim, and Jetter, 2013, for a longer discussion of these tools and how they compare with similar ones). Cross-impact analysis was used first to describe the relationships between each of the descriptors. The team developed a cross-impact matrix that matched each pair of descriptors across all influencing areas. The team determined whether it was plausible that either of two descriptors affected the other, using a three-point scale in which 0 indicated no influence and 3 indicated a strong influence. For example, we determined that the adoption of a national greenhouse gas (GHG)-reduction policy affected the number of vehicles per 1,000 population, but not the other way around. Descriptor pairs that were determined to be totally independent were not carried forward in the analysis; however, they did remain part of the final scenario frameworks.

Second, we applied consistency analysis to those pairs of descriptors in which one was found to influence the other. This analysis examined the various projections for each descriptor. At a sixth workshop attended by RAND, ifmo, and two outside experts, we developed a consistency matrix. Each pair of projections for the two descriptors was rated on a five-point scale, from totally inconsistent to strongly consistent. For example, adoption of a national GHG-reduction policy was deemed consistent with a reduction in the number of vehicles per 1,000 population but not with an increase in the number of vehicles because both vehicle purchase price and driving would become more expensive.

At this workshop, we then fed these results into an online tool called the RAHS platform to group specific projections across all influencing areas. Of all the mathematically possible groupings of projections, RAHS eliminated those that contained total inconsistencies (as defined in the consistency matrix). Of those remaining, RAHS identified groups of descriptors and projections that formed six unique and complementary scenario frameworks. Given that it would be confusing and difficult to develop a multiplicity of scenarios, we deliberately chose to develop only two scenarios. Informed by the discussions at the workshop and visual tools provided by RAHS, we eliminated clusters whose results differed by just a handful of projections. We determined that two clusters represented reasonably different alternative visions of the United States in 2030. These two scenario frameworks became the basis for the scenario narratives.

Produce Scenario Narratives

Drawing on the reasoning and assumptions that surfaced during the expert workshops, we fleshed out the two scenarios into written narratives. These were titled No Free Lunch and Fueled and Freewheeling. To further validate the scenarios, we invited the experts from the workshops, as well as additional experts who had been invited but unable to attend the workshops, to participate in a RAND-developed online Delphi exercise called ExpertLens. Using ExpertLens, the experts rated the scenarios as to whether they were internally consistent, plausible, and understandable, and they provided comments during an online discussion. Although their ratings of the scenarios were generally positive, we used the critical feedback to ensure the relevance and sharpen the content of the scenarios.

Draw Consequences for Future Mobility

The research team estimated the likely outcome of the two scenarios on future mobility. We operationalized future mobility as passenger-miles of travel (PMT) and travel mode shares. We built a matrix that compared each projection in a scenario by four modes: vehicle, transit, domestic air, and intercity rail. For each pair, we rated the directional influence (that is, whether a projection would encourage higher or lower use of a mode), as well as the strength of the influence in each scenario on travel. These estimates were then compared with past trends over the past 20 years to project growth in the use of each mode. From this analysis, the team derived future estimates of PMT and mode shares for different transport modes under both the No Free Lunch and Fueled and Freewheeling scenarios.

Create Wild-Card Scenarios

The research team also developed two wild-card or extremely-low-probability scenarios. Wild cards are designed to provoke "thinking about the unthinkable." These assume that certain events have broken with otherwise-foreseeable trends to move the world in an unanticipated direction. The underlying assumptions of these wild cards originated from comments made at the five expert workshops, in which we asked the experts what events might confound the projections they had just made, as well as from internal discussions at the sixth workshop among RAND, ifmo, and outside experts.

Although multiple possible ideas for wild cards were put forward, we wanted to have wild cards with both negative and positive implications. On the negative side, economic crises were mentioned frequently, likely because of the collective memory of the disruptive nature of the 2008 recession. This became a wild card about a worldwide economic crisis precipitated by events in China, called Red Dusk: China Stumbles. The second wild card was based on comments about the possibility for a disruptive vehicle technology to change mobility in positive ways; this became the Autonomous-Vehicle Revolution.

Why the Scenarios Matter

Because the two scenarios, No Free Lunch and Fueled and Freewheeling, were developed from a systematic, empirical process to identify past trends and prospective projections, they represent plausible futures in which transportation policy and planning might be conducted. These future conditions might be more or less likely and more or less desired. Still, the scenarios provide the opportunity for legislators, public agencies, and private-sector entities to assess and understand how today's decisions might play out in the future.

At one level, the scenarios can provide a valuable reality check on current strategic options and plans. Because the future is uncertain, we do not know whether one, the other, or neither scenario will actually come to be. But organizations can review their strategic plans or policies over the range of futures illustrated by the scenarios to determine whether or not they will be well positioned to address associated challenges and risks. The focus is on the robustness of each strategic option (i.e., can it be delivered in a particular scenario?) and on its importance (i.e., how important is it in influencing a particular scenario outcome?).

Related to this, the scenarios facilitate out-of-the-box thinking. The multiple scenarios encourage people to consider a wider range of futures than in typical day-to-day planning. They enable legislators or strategic planners in organizations to consider atypical opportunities and risks and, by doing so, to identify a more robust set of strategic options.

One of the fundamental uses of the scenarios is to help policymakers and other decisionmakers prepare for change. We encourage individuals and agencies to identify leading indicators of the changes captured by the scenario narratives and to monitor these over time. A leading indicator is typically thought of as an economic indicator that changes before the economy as a whole changes. This concept can and should be transferred to the transportation context. By monitoring leading indicators (or early warning signs) of directions in trends related to each scenario, an agency or organization can explore the question, "Toward which scenario are we moving, and what are the implications of this for our policies or planning?"

Our analysis revealed three driving forces that could lead to one scenario versus the other: the price of oil, the development of environmental regulation, and the amount of highway revenues and expenditures. The price of oil results from the interplay between supply and demand and depends on various determinants outside transportation. In addition, how our scenarios play out will depend on whether and how well transportation policymakers and other decisionmakers anticipate and address upcoming challenges in the other two policy areas.

Report Organization

The remainder of this report is organized in six chapters. Chapter Two provides a short description of past trends in each influencing area. Chapter Three contains the two scenarios. Note that these are written as though we are already in 2030, looking back on developments of roughly the past two decades. For ease of comparing prices, all dollar figures are expressed in constant 2012 dollars so as to avoid making projections about inflation levels. Chapter Four discusses the consequences of the scenarios, including a rough quantification of PMT and explanation of how the driving forces were identified. Chapter Five contains the wild-card scenarios. Chapter Six discusses the potential implications of the scenarios on different levels of government, industry, and private citizens. Finally, Chapter Seven contains our conclusions.

This report also contains seven appendixes. Appendix A describes our methodology in more detail. Appendix B lists the experts who participated in each of the workshops. Appendixes C through G are the white papers, which were developed as background for the workshops. Because of their length, Appendixes C through G are published in a separate, web-only document (Brownell et al., 2013).

Chapter Two
Past Trends in Influencing Areas

In this chapter, we summarize past trends in each of the five influencing areas: demographics, economics, energy, transportation funding and supply, and technology. This information was drawn from the white papers that served as background materials for each workshop. The information provided in these papers, especially the historical quantitative data, helped define the range of plausible future projections. The full white papers, including all references and illustrative tables and figures, are published in Appendixes C through G, available as a separate, web-only document (Brownell et al., 2013). We also discuss briefly the rationale for including each of the influencing areas.

Demographic Trends

Demographics refers to the statistical characteristics of a population. Formal demography is generally limited to basic measures of population size and basic structure and their change over time and space. The research team cast a much broader net in its work. In addition to the basic dimensions of demography—age, gender, and race— we also dealt with *socioeconomic* aspects, such as household size, and *cultural* aspects, such as ethnicity and acculturation. This broader set of concerns is sometimes called *sociodemographics*. This influencing area was included because we know that both the size and the composition of the population influence mobility. Research has found that demand for travel changes with demographic variables, including age, gender, and ethnic and racial group, as well as with population density. Not only do demographic and socioeconomic variables influence travel behavior, but travel behavior, as it manifests in the aggregate, can, in turn, influence the socioeconomic and demographic profile of a city or region. Of the five areas, there was the least disagreement about future demographic projections from the expert panel; many of the projections are based on existing trends that tend to change slowly.

Total Population

The U.S. population has grown from about 180 million in 1960 to just under 310 million in 2010. However, although the United States has added residents every year, the growth rate has generally slowed since 1950. From 1990 to 2010, the population grew at an annual rate of 1.08 percent. (Unless otherwise noted, all population figures are drawn from the U.S. Census.)

Population changes are due to birth rates, death rates, and immigration. Of these, immigration changes the most quickly, with the share of those born outside the United States growing by more than 50 percent in some decades (such as 1990-1999) and negatively in others (2000-2010). Currently, about 13 percent of the U.S. population is foreign-born. The U.S. fertility rate has been between 1.5 and 2.1 births per woman since the early 1970s (Haub, 2009). (A fertility rate of 2.1 percent is considered the replacement rate—that is, the rate at which the total population will remain steady.) Interestingly, the U.S. rate is much higher than that of other developed countries. Finally, life expectancies have increased from about 72 in 1975 (that is, the average person born in 1975 can expect to live to 72 years of age) to nearly 79 in 2010 (Miniño et al., 2011; S. Murphy, Xu, and Kochanek, 2012).

Race/Ethnicity, Age, and Household Size

The racial and ethnic composition of the U.S. population has been shifting in the past 40 years. In 1970, about 83 percent of the population was white. By 2010, that percentage had fallen to 64 percent. Over this same period, the group that grew the most was Hispanics, which increased from less than 5 percent to 16 percent (Ruggles et al., 2010). Growth in the Hispanic population is due to both immigration and higher fertility rates.

The U.S. population is also getting older. The percentage of people over age 65 grew from about 10 percent in 1970 to 13 percent in 2010. In the same time period, the percentage under age 15 decreased from 28 percent to 20 percent (Hobbs and Stoops, 2002; Howden and Meyer, 2011).

Average household size has decreased since 1970, when the average household had 3.11 persons. In 2010, it had 2.63, a slight increase from 2000 (Hobbs and Stoops, 2002; U.S. Census Bureau, 2010a). Even though the proportion of nuclear family households (that is, those with married parents and at least one child under age 18) has declined, the proportions of all other household types have increased: families without children at home, single-person households, and households of unrelated persons (Ruggles et al., 2010).

Urbanization and Vehicle Ownership

The share of Americans living in urbanized areas has increased. In 1970, roughly equal shares lived in central cities, noncentral metropolitan areas (that is, suburban areas), and rural areas. By 2010, about half of Americans lived in suburban areas, one-third in central cities, and one-sixth in rural areas (Hobbs and Stoops, 2002; U.S. Census Bureau, 2010c). At the same time, however, the population density of those central cities has declined by about one-third (from roughly 4,500 persons per square mile to 2,800) primarily because of the outward movement of jurisdictional boundaries.

Vehicle ownership has been increasing as well. In 1970, nearly half of households had one vehicle, while about 35 percent had more than one and 17 percent had no vehicle. By 2010, the share of households with one vehicle had declined to 34 percent. The proportion with two or more vehicles rose to 57 percent, while only 9 percent of households had no vehicle (Davis, Diegel, and Boundy, 2011; U.S. Census Bureau, 2010b).

Economic Trends

Economics is included as an influencing area because the level and growth of economic activity both help determine the amount of travel and are affected by travel. On an individual basis, income and employment are strongly correlated with travel demand. For example, much of the increase in vehicle-miles traveled (VMT) and vehicle ownership in the past 40 years has been attributed to women's increasing participation in the labor force. On a macro level, growth in gross domestic product (GDP) has been identified as a key driver of increased travel. In this section, we explore general economic trends, as well as income, employment, and freight movement.

Economic Growth and Personal Income

Overall economic growth is most commonly measured in GDP. In the United States, the very long-term trend is increased growth; since 1930, the total real U.S. GDP has increased by a factor of 15 (Bureau of Economic Analysis [BEA], 2011). In a recent 20-year period (1991-2011), total real GDP grew at a rate of 2.47 percent annually, or 1.38 percent annually in per capita terms.

Personal income is a measure of Americans' earnings, including wages, income from property, and transfers. Personal income has not grown as quickly as GDP in recent years. From 1990 to 2010, real per capita personal income grew at a rate of 1.07 percent annually (BEA, undated).[1]

Not only has personal income grown more slowly than GDP; the gap between the wealthiest and poorest residents has become larger. The poorest 20 percent of Americans have seen their incomes essentially stagnate in the past 40 years. For the 20th percentile, real household income (that is, adjusted for inflation) was about $18,800 in 1970; by 2010, it was $20,000, an increase of 6 percent. For the wealthiest 5 percent, median household income increased from $114,600 in 1970 to $180,800 in 2010, an increase of 57 percent (U.S. Census Bureau, 2010d). Economists do not agree on the cause of these disparities; some explanations include higher demand for more skilled workers (Bound and Johnson, 1992; Berman, Bound, and Griliches, 1994; Autor, Katz, and Krueger, 1998); global competition, which has led to outsourcing of less skilled jobs (Wood, 1995; Borjas and Ramey, 1995; Feenstra and Hanson, 1996); immigration by unskilled workers (Katz and Murphy, 1992; K. Murphy and Welch, 1992; Borjas, Freeman, and Katz, 1992); and declines in unionization and slow growth of the minimum wage (DiNardo, Fortin, and Lemieux, 1996; Freeman, 1996; Lee, 1999).

Income varies with race, ethnicity, gender, and region of the country. Asian and Pacific Islander families have the highest incomes, at just over $75,000 per year in 2009; black families have the lowest, at $38,000 (U.S. Census Bureau, 2011a). Men earn more than women on both annual and weekly bases, although women's earnings have grown more quickly in the past 40 years (U.S. Census Bureau, 2011b; Bureau of Labor Statistics [BLS], 2011). Finally, per capita personal income is highest in New England, at nearly $49,000, and lowest in the Southeast at $36,000.

----

[1] Estimates of real GDP and real personal income may vary over time because of the use of different deflators in the conversion from nominal to real dollars. As is common, we have used the GDP deflator and Consumer Price Index (CPI) to convert nominal GDP and personal income, respectively, to real dollars. For a comprehensive discussion of the differences between GDP and personal income, see BEA (2007).

Employment

Two long-term employment trends are noteworthy. First, employment by sector has been changing. In 1970, manufacturing was roughly one-quarter of all nonfarm employment; by 2010, it had fallen below 10 percent. Most of the sector growth was in service employment, which increased over the same time period from about 30 to just over 50 percent (BLS, undated).

Second, women's labor-force participation (that is, the percentage of the population over age 16 who either work or are actively seeking work) has risen substantially. In 1980, the percentage of women who participated in the labor force was about 51 percent; by 2010, it was almost 59 percent. During that same time period, the percentage of men participating in the labor force declined, from 77 to 71 percent (U.S. Census Bureau, 2010e).

Freight Movement

Freight movements tend to follow trends in overall economic growth. From 1980 to 2007 (the most recent year for which data are available), freight ton-miles (one ton of freight moved one mile equals one ton-mile) increased at roughly 1.1 percent annually (Bureau of Transportation Statistics [BTS], undated [a]). Most ton-miles are moved via trucking and rail, at approximately 40 percent for each mode.

Energy Trends

Energy is an important influencing area because the cost of gasoline and the availability and cost of alternatives affect the number of miles people drive and the types of vehicles they drive. This influencing area also includes the possibility of legislation or regulations to address climate change. This workshop had the highest number of descriptors, dealing with oil production, consumption, and prices; other sources of energy, including renewables; and policies that address climate change and automotive fuel efficiency. This high number was based in part on the experience in the German workshops, in which energy descriptors were found to be particularly informative in developing the scenarios.

The study focused on eliciting projections related to crude oil. Although the terms *oil* and *petroleum* are sometimes used interchangeably, we use the definition of *crude oil* that is provided by the U.S. Energy Information Administration (EIA): a mixture of hydrocarbons that exists as a liquid in natural underground reservoirs and remains liquid when brought to the surface. *Petroleum* is the broader category that includes both crude oil and petroleum products. We use the word *oil* to mean *crude oil*.

Oil Production, Consumption, and Prices

Domestic oil production has declined in the past 30 years. In 1988, the United States produced enough oil to meet its transportation needs. By 2002, it no longer produced enough oil to meet the needs of the light-duty vehicle fleet (passenger cars and light trucks), let alone other transportation uses (Davis, Diegel, and Boundy, 2011, Figure 1.6 and Table 1.13). In the past few years, U.S. oil production has begun rising again—a result of new discoveries, advances in drilling technology, and the ability to exploit more-challenging resource deposits. However, the share of imported oil remains fairly high, with 2007 imports constituting 58 percent of total consumption (Crane, Goldthau, et al., 2009).

On the contrary, oil consumption has generally increased since 1980, when Americans used about 16.8 million barrels of oil per day, to 2009, when that figure was 19.1 million barrels. However, this trend has not been consistently upward. Consumption fell in the early 1990s, dropping below 15 million barrels per day, and peaked in 2007 at 20.6 barrels per day before declining. Both periods were linked with economic downturns. During both, oil consumption declined across the board—not only transportation but industrial and electric utilities as well (residential and commercial use declined in the 1980s but not after 2007) (Davis, Diegel, and Boundy, 2012).

Oil prices depend heavily on the world market,[2] and retail gasoline prices tend to track oil prices quite closely. In real terms, both oil and gasoline prices were relatively high in 1980 (following the 1970s oil shocks), at about $60 per barrel (2005 dollars) and $3.45 per gallon (2011 dollars). In real terms, both fell through 1990 and began rising substantially only in the mid-2000s. They peaked in 2007 at $87 and $3.45, and prices have been volatile since (EIA, 2011b).

Other Sources of Energy and Vehicle Types

Electric power generation has increased gradually for several decades, from about 2,300 terawatt-hours (tWh) in 1980 to more than 4,000 in 2010. The largest source of electric power is coal, although it has declined slightly from about half of all electricity generated in 1980 to 44 percent in 2010. Generation of nuclear and natural gas has increased during this period, and it makes up the next-largest sources. Of the total generation mix, renewable energy actually declined from about 12.5 percent in 1980 to 10 percent in 2010. Of renewable sources, the largest by far is hydroelectric, but its share of the total declined from about 12 to 6 percent during this time period. The renewable source with the fastest-growing percentage of generation mix in the past few years has been wind, but it accounts for not quite 2 percent of all electricity production (EIA, 2010, Table 8.2a). Retail electricity prices have been rising relatively slowly in the past decade, from $0.066 per kilowatt-hour (kWh) to $0.098 in 2010 (in real 2010 dollars) (EIA, 2012).

---

[2] All oil prices are those for Brent crude, a widely used standard.

AFVs include natural gas vehicles, biofuel or flex-fuel vehicles (ethanol and biodiesel), plug-in hybrid electric vehicles (PHEVs), battery electric vehicles (BEVs), and hydrogen fuel-cell vehicles (FCVs). (Hybrid electric vehicles [HEVs], such as the Toyota Prius, are not considered alternatively fueled because they rely on gasoline or diesel.) Sales of AFVs have been rising rapidly in the United States since the vehicles entered the market in the early 1990s, and, in 2009, it was estimated that between 650,000 and 700,000 were in use. The majority of these run on E85, a mixture of gasoline and up to 85 percent ethanol (Davis, Diegel, and Boundy, 2011, Table 6.1). However, they still constitute an extremely small fraction of the entire vehicle fleet, which is roughly 250 million (BTS, undated [b]).

All-electric vehicles and PHEVs require occasional charging to replenish their batteries. As of April 2012, there were just under 10,000 publicly available charging stations in the United States. (Each charging cable counts as one station, so a physical location with four cables is considered four stations.) (See Office of Energy Efficiency and Renewable Energy [EERE], 2012.)

Energy Policies

Vehicle fuel-economy standards (officially called Corporate Average Fuel Economy [CAFE] standards) became effective in the United States in 1978. All vehicles sold by a manufacturer must meet this standard on average (that is, not every vehicle has to comply, provided the fleet average is met). For passenger cars, the standards rose from 18 miles per gallon (mpg) (11.8 liters per 100 kilometers [L/100 km]) in 1978 to 27.5 mpg (8.5 L/100 km) in 1990. However, they remained at this level through 2011, when the target increased to 30.1 mpg (7.8 L/100 km) (Davis, Diegel, and Boundy, 2012). Because the standards apply only to new vehicles, and light trucks have had lower standards, the average fuel economy of the entire light-duty fleet has risen from about 15 mpg in 1980 to 21 in 2009 (15.7 to 11.2 L/100 km) (see Figure E.4 in Appendix E, Brownell et al., 2013).

The Light-Duty Vehicle Greenhouse Gas Emissions Standards and Corporate Average Fuel Economy Standards Rule Regulations were finalized in 2012. They will begin raising CAFE standards for passenger cars and light trucks to an average of 54.5 mpg (4.3 L/100 km) by 2025 and limit the amount of GHG that vehicles can emit (National Highway Traffic Safety Administration [NHTSA], 2012). Although the U.S. Environmental Protection Agency has been granted the authority to regulate GHGs for vehicles under the existing Clean Air Act laws (42 U.S.C. Chapter 85), legal challenges have delayed the drafting of regulations for other sectors (Moreno and Zalzal, 2012).

Transportation Funding and Supply Trends

This influencing area focuses on the transportation system itself: how it is funded, how much users spend, and whether alternatives, such as mass transit, are available. In terms of funding, we focused on user fees, rather than other taxes and fees that are spent on transportation, because, historically, the United States has had a largely user fee-based system, with a heavy reliance on gasoline tax at the federal level. Projections were made on a per-mile basis for the cost of driving, revenues raised, and expenditures; for road pricing as a new way to pay for the system; and for congestion and transit service.

Initially, the list of descriptors included the total number of road-miles in the country. At the workshop, the experts made a case that, given the very slow change in this descriptor and the fact that much of the U.S. road network is built out, this descriptor would not be very useful for projecting future travel demand. Instead, they proposed using congestion.

Costs, Revenues, and Expenditures

The cost of driving one mile is the average of all fixed costs (such as insurance) and depreciation, as well as variable costs (such as the price of gas) and the fuel efficiency of the vehicle. Depreciation and fuel costs generally account for nearly two-thirds of these costs (American Automobile Association [AAA], 2012). Since 1990, the average cost to drive one mile has risen from about $0.43 to $0.51, in 2008 dollars (Internal Revenue Service [IRS], various years). Much of this increase is due to rising gas prices.

Revenues from user fees, chiefly the gasoline tax, have been declining gradually. The $0.184 federal gasoline tax has not been raised since 1993, and its purchasing power has eroded with inflation. In addition, as vehicles have become more fuel efficient, the amount of fuel purchased per mile traveled has declined. This trend is most pronounced at the federal level; although all states have gasoline taxes, many also use other sources of revenue, such as sales tax, to fund transportation investments. For all levels of government (federal, state, and local), the amount of user fees collected per 1,000 VMT declined from $36 in 1990 to $34 in 2008 (2008 dollars) (see Figure F.6 in Appendix F, Brownell et al., 2013, for details).

Expenditures have increased slightly during this same time period, from $57 to $61 (2008 dollars). This means that, as a proportion of revenue, user fees have declined from about 63 percent of expenditures to 56 percent. Both states and the federal government have filled in this gap with nonuser revenue; at the federal level, the Highway Trust Fund has been kept solvent by transferring revenues from a general fund.

Road Pricing

Road pricing is defined as any form of payment that is tied to direct use of the road network. Tolling, congestion pricing via a zone system, and mileage fees (that is, a fee paid for every mile driven) are all considered road pricing. Some forms of road pricing are quite common in the United States, such as toll roads and managed lanes (that is, lanes that are available to drivers who pay a fee to use them). Others, such as mileage fees and congestion pricing zones, have been implemented only on a limited basis, and seldom in the United States. (Several states charge a type of mileage fee for trucks, and several states have conducted pilot studies on, but not implemented, mileage fees. Some cities, such as San Francisco and New York, have considered but not implemented zone-based congestion pricing.)

Transit Service

The provision of transit service, as measured in transit vehicle-miles (that is, the number of miles traveled by transit vehicles in revenue service, which is the operating period in which passengers can board and ride on the vehicle), has been steadily increasing in the past few decades. In 1990, transit operators collectively provided 2,917 million miles of transit service. By 2010, this increased to 3,530, an annualized increase of 1 percent (American Public Transportation Association [APTA], 2012). This is a national figure; the changes vary between metropolitan areas.

Transit service quality (as opposed to quantity) is more difficult to measure because of its subjective nature. However, transit agencies do assess the conditions of their vehicles (both rail and bus) on a scale of 1 to 5, with 1 being the worst (seriously damaged) and 5 the best. On average, across transit agencies, the condition of the rail fleet declined from 3.9 to 3.3, while the bus fleet improved from 3 to 3.2 (Federal Highway Administration [FHWA], 2003, Exhibits 3-38 and 3-51; FHWA, 2010, Exhibits 3-24 and 3-26).

Congestion

Because this descriptor was added at the workshop itself, we had not previously developed any data on past congestion levels. However, growing congestion is a critical issue in the United States. The Transportation Research Board (TRB) has called the early 21st century the era of congestion (Executive Committee, 2009). For example, yearly hours of delay per auto commuter in large urban areas, on average, have almost tripled from 19 hours in 1982 to 52 hours in 2010. Even in medium-sized urban areas, average yearly hours of delay tripled during the same period from seven to 21 hours (Shrank and Lomax, 2012). Although congestion has been quantified in various ways, defining a quantitative descriptor on the fly during the workshop would have been challenging and time-consuming. So, at the workshop, we agreed on a qualitative descriptor, level of congestion. This encompasses both the number of hours per day when road conditions are congested and the number of congested locations.

Technology Trends

The rapidly changing nature of technologies related to driving and telecommunications seems very likely to be a major influence on mobility during the next two decades. However, unlike the other four influencing areas, there is relatively little past experience in these areas on which to draw. Therefore, the projections developed about technology are, by definition, more speculative than the others. At the workshops, the descriptors included Internet access and use, in-vehicle technologies, and data privacy.

Internet Access and Use

Internet use has the potential to affect mobility in several ways. It might substitute for travel (for example, if a shopping trip to purchase a book is replaced by a download of an e-book), modify travel (for example, if online mapping provides a different route to a destination), spur additional travel (for example, if learning about a new restaurant online prompts a trip to have dinner), or have no overall effect. Access to the Internet has followed a typical S-curve since 1990; as of 2010, about 80 percent of Americans had Internet access (World Bank, 2011). Of these, almost all have broadband access (Economics and Statistics Administration [ESA] and National Telecommunications and Information Administration [NTIA], 2011).

One major substitution that depends on Internet access is *telework*—work performed at home or from another location close to home that would otherwise be performed at a workplace. No single survey reliably tracks the percentage of Americans who telework, and varying definitions make it difficult to compare figures across surveys.[3]

Another substitution is shopping trips. Online retail sales remain a very small fraction of all retail sales—just 4 percent of the total in 2009, according to the most-recent data from the U.S. Census Bureau (2011c, Table 1055). A different survey measured the number of adults who had ever made an online purchase; this figure rose from 22 percent in 2000 to 49 percent in 2007 (Horrigan, 2008).

---

[3] Surveys vary in how they ask about the frequency of telework, in how they define *telework*, and in whom they ask, which makes finding statistics for which a trend can be discerned a challenge. Some surveys ask only about the primary mode of travel, while others ask about whether any days during a specific period involved telework. Surveys do not always distinguish between working at home, commuting, shifting commute times out of peak periods, and working from home outside regular business hours. As a case in point, the U.S. Census Bureau conducts two surveys that report on home-based workers, the Survey of Income and Program Participation (SIPP) and the American Community Survey (ACS), and each uses different definitions. The SIPP reported that, in 2010, 13.4 million of 142 million workers (9 percent) worked at least one day entirely at home per week. In 1997, it was 9.2 million of 132 million workers (7 percent). To be regarded as an at-home worker by this survey, a respondent age 15 or older must report having worked only at home on a given workday. Individuals who check email or carry out other work activities at home but outside of normal work hours are not counted as home-based workers. On the other hand, the 2010 ACS reported that 5.8 million workers (4.3 percent) usually worked at home during the week before the interview. In 2005, this was 4.8 million workers (3.6 percent). The ACS used a different definition: how workers age 16 and over usually got to work in the preceding week. For those who used several modes, respondents were asked to list the mode used most often. The ACS data assume that respondents who select work at home presumably work the majority of the week from home. So, although these two data sources report on very different realities, one trend on which they seem to agree is that working at home is on the rise. At the workshop, we presented data indicating that the percentage of working Americans who telecommute as their usual means of working rose from 3.5 percent in 2005 to almost 4.5 percent in 2010. However, in compiling this report, we have been unable to confirm the provenance of this figure.

In-Vehicle Technologies

In-vehicle technologies include several technologies: telematics, or the integration of communications with trans-portation technologies; advanced driver-assistance systems (ADASs); and autonomous (self-driving) vehicles. All of these are, of course, fairly new, so information on past trends is not particularly meaningful for extrapolation.

Telematics includes relatively widespread applications, such as global positioning systems (GPSs), as well as fairly recent applications, such as real-time traffic information. We were unable to locate any reliable information about the percentage of the current vehicle fleet equipped with these technologies, let alone any past information. Similarly, we were unable to locate reliable information about the market penetration of ADASs, which include such technologies as crash-warning systems, adaptive cruise control, and automated parking. Autonomous vehicles have been tested, and three states allow them to be operated on public roads for test purposes, but they are not available commercially as of this writing.

Data Privacy

In-vehicle technologies, although they may become widespread and provide enormous benefits to both individual drivers and the transportation system as a whole, raise concerns about the privacy of drivers' personal informa-tion. Although we cannot make specific quantifiable projections about data privacy, the current situation with regard to data privacy helps explain why this issue is so controversial. The United States does not have a unified approach to privacy, and no single law or regulation governs privacy at the federal level. Important issues around the use of devices, such as GPS units, and whether the data they generate are considered private have yet to be fully determined in the courts. In addition, although research suggests that people strongly value privacy, it also shows that people have inconsistent attitudes depending on which technology is collecting their data (Nguyen and Hayes, 2010).

Demographic Trends

Technology Trends

Transportation Funding and Supply Trends

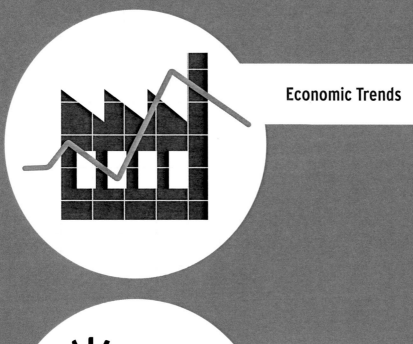

Economic Trends

Energy Trends

Figure 2.1. Past Trends in Influencing Areas

Chapter Three
The Scenarios

In this chapter, we present the scenarios, No Free Lunch and Fueled and Freewheeling. However, we preface these scenarios with a section on common projections. These are estimates of future possibilities that are shared by both scenarios. In keeping with common practice in this field, the content in this chapter is written from the future vantage point of 2030. Note that all prices are provided in 2012 dollars.

Common Projections

Although the overall focus of scenario planning is identifying alternative future developments, different scenarios often share some common projections. This can happen because there is only one (relatively certain) projection for a specific descriptor by the expert panels or because the cluster analysis that underpins the scenario-building process determined that one projection would fit both scenarios. This section presents those projections that are common to both scenarios. Because their probability can be regarded as rather high, the value of these factors is significant.

Demographics

The population has continued to grow from 2010 to 2030 at 0.8 percent annually. By 2030, the total population is about 360 million, compared with just under 310 million in 2010. This growth—which is quite different from that in other developed countries, where populations are stabilizing and, in some cases, shrinking—has several causes. Fertility rates among American women are at replacement level, and immigration has also grown. Immigration rates have increased, from 1.5 percent in the 2010s (lower than previous rates that were suppressed by the recession in 2008) to 2.5 percent in the 2020s. Immigrants currently make up about 17-18 percent of the total population.

The population has also grown noticeably grayer and browner. The proportion of Americans over 65 has risen from 13 percent in 2012 to 18 percent in 2030. The majority of these older citizens (almost three-quarters) are white. However, the younger generations are far more racially and ethnically heterogeneous. As a result of both Latino immigration and higher fertility rates in that segment, the proportion of all Americans who are Latino has grown from 16 percent in 2010 to between 20 and 25 percent. The absolute number of whites, who now make up about 58 percent of the total population, has not declined, but the population increase has been in nonwhite groups.

The average household size is currently 2.6 persons per household. Although this is unchanged from 2012, it masks two opposing trends that have been occurring. On the one hand, the number of one-person households has increased, mostly because of older people living alone. On the other hand, the share of larger families (more than two children) has also increased, which is due, in large part, to immigrant families having higher-than-average fertility.

Economic Growth

From 1991 to 2001, growth (as measured in total, not per capita, GDP) was, on average, 2.47 percent per year. Although both scenarios posit that economic growth picked up strongly after the 2008 financial crisis, the average annual growth rates over the past two decades differ somewhat. In the No Free Lunch scenario, growth has averaged 2 percent per year, while, in Fueled and Freewheeling, growth has been even stronger, at 2.5 percent. The difference between the scenarios is due to the higher price of oil assumed in No Free Lunch, which depressed exports to some extent. Personal incomes also vary with GDP growth rates. In No Free Lunch, they grew by roughly 1.2 percent per year, while, in Fueled and Freewheeling, they grew by closer to 2 percent.

Freight movement followed these trends because it remained linked to changes in total GDP. From 1980 to 2007, ton-miles grew at about 1.1 percent per year. In these scenarios, growth rates are 0.9 and 1.1 percent, respectively (roughly 45 percent of total GDP growth, a ratio similar to that of the 1990s and 2000s).

Nevertheless, in terms of economic development, the scenarios have more similarities than differences. In both, income inequality has increased from its already high degree in 2012. Factors explaining this development include inequality in access to education because of unaffordable tuition, the tax system having fostered greater concentration of wealth, and income stratification having become more pronounced between high- and low-skilled labor.

In addition, employment trends have continued shifting away from manufacturing, with 2030 employment in that sector at about 7–8 percent, down slightly from 10 percent in 2010. Service employment, in contrast, continued to increase, from just over half in 2010 to 55–60 percent in 2030.

Domestic Oil Production

Although U.S. total oil consumption in the two scenarios is different, domestic production has increased quite substantially in both cases, from 8 million barrels per day in 2010 to 15 million barrels per day in 2030. This is largely due to the expensive, large-scale ramp up of new production facilities for unconventional oil and gas sources, which proved to be more productive than anticipated. For example, tight oil production (in an area stretching from Texas to the Northwest) increased, adding 3 million barrels per day. Oil sands also increased in production, following Canada's lead (Canada was already successful in producing oil from oil sands in the 2010s). Oil shale was more speculative and did not increase by the same numbers as tight oil because of environmental opposition. Regardless of U.S. demand, which varies between the two scenarios, strong demand from rapidly developing countries means that new oil production could easily be absorbed on the world market.

Technology

The development of advanced consumer and automotive technologies has been strong in some areas and less so in others. Household access to digital information using broadband technology has increased from 68 percent in 2010 to 90 percent in 2030. In turn, this has helped online retail sales grow to 30 percent of all transactions (not dollars), up from an estimated 4 percent in 2009. Telematics in vehicles have become fairly widespread, with about 60 percent of all vehicles having some type of advanced telematics (such as GPS) (there were no reliable data to measure this in 2012). Many of these advanced systems will be through vehicle-device interface, rather than exclusively built into the vehicle. ADASs are also widespread, with about 90 percent of new vehicles purchased having one or more of the options available. However, autonomous vehicles, which had just entered the public consciousness in 2012, remained scarce, with no more than 5 percent of vehicles operating with a partly or fully automated driving capability.

Corporate Average Fuel Economy Standards

CAFE standards, which set an average target for fuel efficiency in miles per gallon, were adopted in the United States beginning in the 1970s. In 2012, Congress passed new standards mandating that, on average, all vehicles produced by 2025 get 54.5 mpg. This average applies to all passenger vehicles, including cars, light-duty trucks, and medium-duty passenger vehicles. This represented a large increase from the 2012 average standard of 30.1 mpg. The standards were phased in gradually, rising slightly each year beginning in 2013.

Scenario 1: No Free Lunch

Overview

By 2030, several factors have combined to bring about strengthened regulations to reduce dependency on oil and GHG emissions. Oil prices have soared to $190 per barrel based on Chinese and Indian demand, as well as political instability in the Middle East. But perhaps more importantly, the destructive effects of climate change have become increasingly clear to the average American. Floods and droughts have led, over the past decade, to skyrocketing food prices and insurance losses, and, slowly but surely, the general public began to pressure elected officials to take action. The United States introduced a cap-and-trade program in 2022, supported investments in alternative fuels and vehicles, and developed widespread road pricing programs. Despite some grim predictions of economic demise, the investments have spurred economic development, and the U.S. economy has continued to grow at a healthy pace. The key drivers and how they shape this scenario are outlined in Figure 3.1.

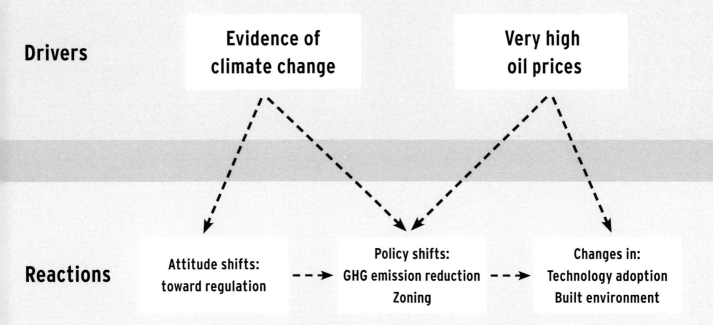

Drivers

Evidence of
climate change

Very high
oil prices

Reactions

Attitude shifts:
toward regulation

Policy shifts:
GHG emission reduction
Zoning

Changes in:
Technology adoption
Built environment

Figure 3.1. Key Drivers and Reactions, No Free Lunch

Oil Prices, Rising for Years, Hit an All-Time High

Earlier in 2030, oil prices hit an all-time high of $190 per barrel (Brent crude). This peak did not reflect a straight-line increase from the spike in 2007-2008; prices were extremely volatile in the 2010s and 2020s. Pain at the pump is a fact of life. In the past two decades, some years have been merely uncomfortable price-wise—no higher than $5 per gallon. In others, supplies have dwindled, causing long lines at the pump. This was a real shock; Americans had not experienced extreme supply constraints since the early 1970s, when the retail purchase of gasoline was limited to vehicles with odd-numbered license plates on odd-numbered days, and vice versa.

The high price of oil, unimaginable 20 years ago, has three major causes. First, sustained and significant instability in the Middle East has meant supply disruptions, leading to the price spikes as other producers scrambled to keep up with demand. Second, the $190-per-barrel price reflects strong demand from China and India. In 2009, China surpassed the United States as the largest car market in the world, and India's market has been expanding rapidly. Because of the sheer size of their populations, neither appears to be anywhere near saturation. Third, environmental regulations have made some types of energy production more expensive, which has constrained the supply to some extent.

Some observers thought that prices might be mitigated by new sources of oil, especially as the Arctic opened up to drilling as sea ice continued to melt rapidly or by new technologies. However, early estimates about the ease and cost of extracting oil and natural gas with hydraulic fracturing proved to be overly optimistic. Well productivity fell more sharply than expected, regulations added to the cost of extraction, and, after bitter disputes over contaminated water supplies, some local governments banned the use of hydraulic fracturing altogether. In another example, the process of converting coal (which remains plentiful) or natural gas to liquid fuel is too expensive to be profitable because of climate regulations.

This peak is not only unprecedented; it has shaken the confidence of the public and elected officials alike. Gas prices of $8 to $10 per gallon have changed both the kind of cars Americans want to buy and how much they drive them. The average increases in oil price have outpaced fuel-economy improvements for conventionally fueled vehicles, which really focused attention on reducing America's dependence on oil.

A Changing Climate Sparked Changing Attitudes

Although scientists had been sounding alarms for years, in the 2000s, the American public had an ambivalent attitude toward climate change. Awareness rose, and, for a short time, Congress considered taking action, but skepticism also rose, and many elected officials cast doubt very publicly on the scientific basis for believing that human-caused carbon emissions create climate change. However, these attitudes began changing in the mid-2010s, when several weather-related disasters in rapid succession could no longer be brushed off as mere coincidence.

There was the Midwest "flash drought" of the early 2010s that severely negatively affected the country's corn, soybean, and livestock industries, leading to unprecedented increases in the costs of staple foods. "Superstorm Sandy" flooded coastal areas of New York and New Jersey in the fall of 2012. Intense "derecho" storms (widespread, long-lived, straight-line windstorms associated with fast-moving severe thunderstorms) in the summer of 2015 caused month-long power outages across the East Coast. Large tracts of arid land in the Southwest burned in uncontained wildfires in 2017–2018. Then, after prolonged drought in the Great Plains, many rural Americans in the late 2010s who relied on wells for water had none. Wells across the region had run dry. Although climate skeptics remained, they were few and far between by 2020. Once people saw the effects on their own lives, public attitudes changed quickly.

But even public attitudes might not have been enough to effect legislative change had they not been accompanied by changes in the business community. From 2010 to 2019, the United States recorded more billion-dollar disasters than it had during the preceding four decades, which affected the home insurance market. Some insurance companies began to stop writing policies in areas deemed prone to floods and fires, while others increased premiums for all policyholders. When insurance premiums rose, even in safer areas, homeowners were outraged to be subsidizing people who insisted on living in floodplains. Events from water shortages to floods were disrupting the supply chains for Fortune 100 companies. Corporate America came to realize that the impacts of climate-change events could not be deflected from their bottom lines.

National Greenhouse Gas-Reduction Policies Implemented

In 2022, the United States passed legislation (popularly known as "Green Cap") for a national emission trading scheme (ETS) that uses emission trading as the primary vehicle to drive carbon pollution reduction. A decade before, such sweeping legislation would have been surprising, to say the least, but the political climate had changed along with the global climate. The issue emerged as a major one in the presidential election of 2020, and the challenger made climate legislation a key campaign promise. The voices in favor of enacting comprehensive legislation had grown louder, including persuasive ones in the chambers of commerce.

The business community encouraged regulation based on its perceptions of risk and lobbied for regulatory certainty. It joined forces with environmental advocates to push legislation that, although not perfect, constituted a significant change. It helped, too, that the United States had climbed out of the 2008–2010 recession and the economy was growing at about 2 percent annually.

The legislation covers the largest emitters of GHGs—big power plants, petroleum refineries, paper mills, and other industries. Although the ground transportation sector is not specifically called out in the higher CAFE standards, the standards are seen as sufficient incentive for reducing energy use in that sector. Since passage of the CAFE standards, the government has managed to set emission caps and reduction targets for each trading period without much partisan bickering. In 2025, the aviation sector was included in an open ETS consistent with the United Nations Framework Convention on Climate Change (UNFCCC).

Legislation Spurred Innovation and Restructuring

Since 2022, the price on carbon has created an incentive for the big polluters to use or generate renewable energy, reduce energy consumption, implement technologies that will improve energy efficiency, and invest in renewable energy. Now wind power or solar power plants exist in areas that were previously the exclusive province of heavyweight U.S. polluters, such as the coal-producing regions. Renewable portfolio standards (state mandates that the fuel mix used in that state contain a minimum amount of renewable energy sources) and clean-energy standards have been widely implemented and are being enforced. Economic growth has also furthered investment in renewables.

Earlier this year, EIA announced that total U.S. carbon dioxide (CO_2) emissions for 2029 fell dramatically. This decrease has been attributed primarily to greater use of renewable energy. The total share of nonhydro renewables in the power generation mix in 2030 has doubled to 20 percent in the past two decades. Green Cap also helped stimulate a significant shift from coal to low-emission sources. The share of coal in electrical generation is 30 percent, having decreased from 45 percent in 2010. However, U.S. coal is still being shipped in greater amounts to China, the world's largest consumer of coal. Virtually all these changes can be attributed to Green Cap. Prior to that, industry did not have the incentives or mindset to pursue renewables more aggressively.

Natural gas prices are lower than oil prices, sustained by consistent production from shale resources. Despite these lower prices, no gas-to-liquids facilities have been constructed, and there are limited natural gas exports, due to both the Green Cap policy and political pressure from industry and electric utilities that have predominantly shifted to natural gas use. Heavy-duty trucks have adopted liquefied natural gas as their primary fuel, and many bus fleets are fueled by compressed natural gas. Yet, due to Green Cap, as well as decisions about electricity charging infrastructure deployment and the relatively low cost of batteries, fully electric vehicles are more popular than compressed-natural gas passenger vehicles.

Some of the revenue from the carbon trading scheme has been directed to alternative fuel research and development (R&D), leading to significant expansion of the AFV fleet in the United States. In the 2020s, adoption of AFVs by American consumers has been strong (about 30 percent of all new light-duty vehicles sold in 2030) as both higher oil prices and regulations have increased the cost of driving conventional vehicles. The cost to drive has doubled in the past 20 years, reaching $1.04 per mile in 2030, with the greatest share of the increase in the past ten years. (At the current average CAFE standard of 54.5 mpg, adopted back in 2012, an average gas price of $8 per gallon means that gas costs about $0.14 per mile, roughly the same as in 2012. But new safety features and the materials and technology to meet the standards have made new cars more expensive, prompting drivers to keep their vehicles longer. Therefore, many people with older cars pay the equivalent of $0.25 per mile in fuel costs.)

The most-popular vehicles in the AFV fleet are electric vehicles (EVs), helped along by relatively cheap electricity and innovative battery technology. This, in turn, stimulated the development of a critical mass of public charging infrastructure, but, with time, it became clear that people preferred to charge at home or work. In 2030, about 85 percent of EV charging is done at home and 10-15 percent at the workplace. A battery charging station has become standard in most new homes, and retrofitting older homes has created hundreds of new small businesses. Most employers provide free or low-cost charging to employees, similar to the perk of providing free parking. The widespread availability of workplace charging has become an added incentive for consumers to purchase an EV. Now the investment in public infrastructure is seen as a "safety net" to EV owners, who are mostly charging at home anyway.

The air transport sector has also experienced various changes in the past two decades. Since the aviation sector became a net buyer of emission certificates, most U.S. airlines have had to increase fares to cover the high fuel prices and additional carbon costs. Even before 2025, a large number of airline consolidations had taken place worldwide. Surviving airlines increased their average aircraft size, invested in fuel-efficient aircraft, consolidated their networks, and decreased frequencies on certain routes to maximize profits. Complex market dynamics created an environment in which airlines generated more profit transporting fewer passengers at higher airfares than more passengers at lower airfares, reducing both economic and geographic access to the system, with worldwide repercussions for tourism and business development. All in all, the air transport sector kept growing over the past 20 years but at lower rates than previously expected.

Gross Domestic Product and Oil Consumption No Longer Coupled

Although some predicted that the additional regulatory burdens of meeting Green Cap would slow the recovery, the opposite has proven true. High oil prices and the costs of meeting Green Cap may have hurt some industries, but they have also spurred a host of new innovation, and, in turn, the new investments have fostered economic growth, creating jobs for scientists and researchers, as well as blue-collar jobs in manufacturing and installing everything from solar panels to home charging stations. Both before and after Green Cap, the United States experienced sustained moderate economic growth, averaging about 2 percent per year.

Personal incomes have increased only about 1.2 percent annually for the average person. Many more people have discretionary money to spend. Unfortunately, income inequality has also continued to increase because of more-pronounced income stratification between high- and low-skilled labor and unequal access to education for poor people. Furthermore, climate-change events have affected regions of the country differently. The coasts are still doing better than the Midwest, although rising sea levels on the coasts have affected low-lying areas whose economies were based mainly on tourism.

Productivity has continued to grow about 1.6 percent annually. Labor productivity increased because of such factors as technology advances (much spawned by Green Cap) and the availability of a highly skilled labor force. The agriculture sector has done well because of increasing food demand from Brazil, Russia, India, and China (the BRIC countries), despite the changing agricultural patterns associated with climate-change events. Coupled with revenues from emission trading and road pricing, transportation infrastructure investment has increased.

Freight ton-miles have increased annually by 0.9 percent, consistent with the GDP growth. But mode shares have shifted. Since about 2018, most freight has been carried by rail. Waterway transport has also increased significantly. Transportation costs for shipping by rail and waterway have fallen because freight ton-miles by these modes have risen as a share of all miles. The trucking industry has been victim to volatile oil prices, and shippers got tired of the radically different prices that were being charged for same distances and same loads in different years and different times of the year. Air-freight demand had been hampered by high fuel costs and emission levies that have caused operating costs to skyrocket.

Oil Consumption Down, Supply Up

Between the spike in oil prices and the adoption of AFVs in many transportation sectors, the United States reduced its oil consumption to 16 million barrels of oil per day in 2030. This is a substantial decrease from 19 million per day in 2009. Demand for gasoline had peaked before 2012, and oil-intensive industries have become smaller and more efficient because of Green Cap. Heavy-duty trucks have adopted liquefied natural gas as their primary fuel. Marine fuels have been replaced by natural gas. These changes in fuel type helped the marine industry to compete, but not so much the trucking industry. Not only had CAFE standards been adopted but states also adopted renewable portfolio standards and, in some cases, low-carbon fuel standards. All this meant that the United States was able to meet standards under the federal Renewable Fuel Standard (RFS) program, which mandates increased use of renewable fuels (produced from crops, such as lignocellulose, an inedible plant by-product) blended into the gasoline supply. Adoption of these new policies was relatively painless because of shifting public and corporate attitudes regarding the utility of government regulations for the common good.

Although consumption is down, the United States has been increasing its production of oil to reduce its dependency on imported oil, especially from the Middle East. In 2030, the United States is producing 15 million barrels per day, nearly double what it was producing in 2010. The high price of oil and strong demand in India and China had provided important incentives to seek out unconventional sources. These sources have proven to be more and more productive. Starting in about 2013, tight oil production increased in an area stretching from Texas to the Northwest, and then a bit later, from 2015 onward, oil shale and oil sands production increased as well. However, as noted above, these sources remained expensive to exploit under Green Cap, so they still constitute a relatively small proportion of production.

Urban and Suburban "Densification" Increased Public Transit Ridership

Immigrants and young people have continued to settle in urban and suburban areas. Zoning regulations to arrest suburban sprawl have led to an increase in population density in these areas. New freestanding homes are smaller and closer together, and more row houses and apartments are built than previously. Housing is more expensive because developable land is scarcer, and land prices per square foot have increased.

A variety of factors have increased transit demand in the past two decades: higher densities, a doubling of the cost of driving due to high oil prices and road pricing programs, and increasing congestion. In most metro areas, transit agencies have responded by adding about 35 percent more miles of service than they provided in 2012. Although the fuel economy of conventional vehicles has improved considerably, the cost to drive is still high enough to discourage drive-alone commuting. Increasing demand has led to increased funding for public transit, from both ridership revenue and road user fees. Users and potential users have seen improvements in the quality of public transit as well, which has further increased ridership, especially among choice riders (riders who have a choice about whether to use transit or to drive). Increased transit options and increased cost to drive have led to reduced vehicle ownership in areas well-served by transit.

Of course, transit works only for certain trips, and much of the increased demand has been for commuting. So some, particularly younger people, have sought alternatives to mobility, leading to substantial upsurges in telecommuting. In 2030, 40 percent of workers telecommute, meaning that, on any given day, 40 percent of employees are working either at home or from another location close to home that is not their usual work-place.[4] Similarly, online shopping has grown to 30 percent of all retail sales, in dollars, from about 4 percent in 2009. Trips that cannot be substituted with telework or other virtual communications are often taken in shared vehicles. Car sharing has benefited from innovations in social networking technologies, such as peer-to-peer car sharing, in which one person can easily lend his or her vehicle to online contacts. Car sharing is a usual mode of travel, even in some suburban areas.

"Aging in Place" Creates Diverse Mobility Needs and Opportunities

The mobility culture of the older generation, who grew up in the 1950s and 1960s, values the freedom afforded by personal automobiles. The "young elderly" (age 65–75) have continued to take work, shopping, and leisure trips by vehicle to the extent that they can afford, given the high cost of fuel and prevalence of road pricing. The share of new vehicles that have been equipped with ADASs has increased from 20 years ago, when only premium vehicles were equipped with ADASs, to nearly 90 percent of all new vehicles today. Some of this rapid adoption is due to standards introduced in 2013 by NHTSA for connected vehicles. In addition, the young elderly have been a strong market. They purchase ADAS-equipped vehicles to drive longer and qualify for the steep discounts given by insurance companies. These systems keep cars from running off the road and from colliding; they also help with parking. Consequently, VMT among the young elderly have been reduced only modestly. Travel by public transit has not become a preferred option for this age segment. But many of them have bought AFVs to mitigate the high cost of oil and to benefit from the AFV purchase tax credit introduced under Green Cap.

The "older elderly" (75+) have sought alternatives to driving and taking public transit. In the 2010s, the number of paratransit services (both volunteer and for-profit) increased to transport the older elderly to medical and other destinations, given their large numbers in many suburban communities. Although systems are not yet market-ready, entrepreneurs have been heavily involved in R&D efforts for autonomous vehicles to populate fleets of taxi-like services to serve the transportation needs of the elderly. The technologies developed in the early 2010s did not work out well for liability and other reasons, but the R&D efforts of entrepreneurs have been paying off in the 2020s. The share of autonomous vehicles (both partly and fully autonomous) remains very low (not more than 5 percent) in 2030, but the expectation has been that it will expand significantly in the 2030s and 2040s.

---

[4] In the workshop, the experts' projections ranged from 15 to 45 percent. In the course of the workshop discussion, the experts settled on a high projection of 40 percent and a low of 15 percent. We have kept these projections because they reflected the expert opinion. We recognize that the assumption that 40 percent of the workforce will greatly reduce or eliminate their commute by telecommuting on any given workday may be unrealistically high. The expert participants were clear that their definition was not a person's "usual" place of work. Since the workshop, we have not been able to identify data support for the projection that 40 percent of workers will be telecommuting on any given day by 2030 that includes a specification of the survey population or definition of telecommuting. We reran the RAHS analysis with projections of 20 percent and 30 percent. Neither of these lower projections leads to a different scoring on the consistency matrix. And rerunning the analysis did not change the RAHS results, in terms of either the raw scenarios or the clustering. The use of this projection and others in the analysis and subsequent scenario development is described in Appendix A.

Road Pricing Goes Mainstream, Providing Essential New Revenues

Congestion has increased naturally with the increase in population growth and in ton-miles of freight. No (net) new roadways were built in the 2010s; however, technology solutions had helped to better manage existing road networks. But still, the United States had underspent on roadways for so long that eventually the country had to make up for it.

Over time, the gasoline tax has increased only slightly, despite a few attempts to increase it more substantially as part of deficit negotiations. The most significant contributor to the availability of transportation funding in 2030 has been that congestion pricing was introduced in several large metro areas in the early 2020s. Because elected officials and transportation planners observed that congestion pricing both raised revenues and helped to manage system performance and that gasoline taxes could never totally provide the needed revenues to maintain and expand the surface transportation system, they made stronger and more-public cases about adopting road pricing. As the public began seeing reductions in congestion, along with better-maintained roads, resistance lessened. Also, concerns about privacy (or black-box technology in vehicles) were allayed through appropriate and effective data privacy standards.

With the relative success of congestion pricing, other forms of road pricing became more common by 2030. Priced lanes and variable toll roads crisscross every state and metro area, with the result that the country now raises about $45 (in user fees of all types) per 1,000 VMT, a 30-percent increase since 2008. As a result of these new revenues, the United States can collectively spend about $80 per 1,000 VMT, also a 30-percent increase. Some localities have also introduced variable parking pricing and mileage-based user fee systems. User fees were introduced for AFVs once these reached a high-enough share of the vehicle fleet. The federal government even allowed tolling of the interstate system in rural and urban areas. As the prevalence of tolling increased, providing new revenues, surplus revenues were transferred to expand and maintain public transportation.

Opportunities and Challenges Ahead

The United States seems to be on sound footing in 2030. It is dealing with the issue of climate change directly, having reached a national consensus on its causes and effects, and, to the surprise of many, it has had a positive rather than negative effect on the economy. We have lowered our dependence on oil and, through road pricing, put the transportation system on a more stable financial footing. Dealing with the effects of climate change that have already taken place remains a challenge, but the revenues available (through taxes and fees) to address those communities hit hard by drought and flooding has made this a more manageable issue. However, both high taxes and high oil prices may lead to slowed economic growth in the future, including difficulties competing internationally against countries with lower tax rates.

Scenario 2: Fueled and Freewheeling

Overview

Looking back from 2030, the situation seems similar in many ways to that of the 1980s and 1990s. Lower-than-anticipated oil prices for the past 15 years have led to a world quite different from that predicted during the financial crisis in 2008. The economy is booming, the anticipated harms from climate change have not unduly affected the United States, and inexpensive fuel and electricity have made driving relatively inexpensive. On the negative side, congestion is increasing, and the country as a whole is not making needed investments in infra-structure because of reduced funding for transportation. Many caution that this period of inexpensive oil has to be nearing an end soon, but, for now, the governing assumption is that major change is not needed to address that future. Figure 3.2 shows the key drivers and how they shape this scenario.

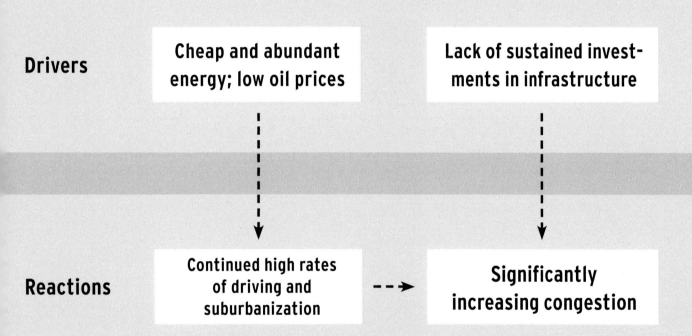

Drivers

Cheap and abundant energy; low oil prices

Lack of sustained invest-ments in infrastructure

Reactions

Continued high rates of driving and suburbanization

Significantly increasing congestion

Figure 3.2. Key Drivers and Reactions, Fueled and Freewheeling

Missed Opportunities to Change Course on Energy Policy

When we look back on the past few decades, history seems to have repeated itself. Twice, intertwined energy and financial crises set the stage for what might have been a broad change in the country's direction, and yet, when the crisis abated, the country remained on roughly the same course. Both of these might have been turning points in energy policy but ultimately produced little change. The first potential turning point came in the 1970s, and the second in the late 2000s.

The twin oil shocks in 1973 (the Arab oil embargo) and 1979 (the sudden loss of production in Iran) suddenly vaulted energy conservation, not to mention energy independence, to the top of the national policy agenda. Then-President Jimmy Carter put solar panels on the roof of the White House, and Congress enacted CAFE standards to ensure a minimum fuel economy in new vehicles. Long lines at gasoline stations and the unnerving prospect of gasoline rationing suggested that the United States might take serious action to wean itself from foreign oil and maybe even away from oil more broadly.

In the 1980s, the country abruptly changed course as oil prices fell dramatically. The solar panels came down, and sport utility vehicles came roaring onto the roads. The country started booming economically, albeit with a few fits and starts, through the dot-com era in the late 1990s and the 2001 terrorist attacks that upended Americans' concept of security. For roughly 30 years—from the 1980s through the 2000s—U.S. energy policy generally assumed continued dependence on oil, and CAFE standards remained unchanged.

The second potential turning point came in the late 2000s. The financial crisis of 2008 caused the deepest recession (often called the Great Recession) since the Great Depression, with significant unemployment and upheaval in the housing sector as prices fell. Oil prices spiked, and people talked more about climate change, especially after the droughts of the early 2010s reduced the corn harvest, pushing up food prices in some regions.

Then once again, as the immediate crises abated, the country resumed its previous course in the mid-2010s. The Great Recession turned out not to be a major inflection point but an interruption. Although some observers thought that it carried warning signs—the debt was too large, banking regulations were too lax, the indicators of climate change were too great to ignore—whether for better or worse, Americans were all too relieved to put those crises behind them, especially as the economy picked up again.

Oil Prices Stabilized as the Economy Pulled Out of the Great Recession

A key factor in the prosperity of 2030 is the current price of oil: $90 per barrel, or lower than it was in 2012 (about $99—both prices for Brent crude). In 2014, one of the major oil producers, which had been quietly sniffing around Alaska, announced that it had struck gold, at least figuratively speaking. A major underground deposit was suddenly available for drilling, and all the necessary factors were in place to exploit it. The technology had come down sufficiently in price to make it worthwhile to pursue, and weak environmental regulations allowed drilling to proceed without extensive roadblocks or controversy. Canada had already made it easy for tar sands to be developed, and the opening of the Keystone XL pipeline in 2016 provided an inexpensive way to bring crude oil down from Alaska to refineries in the rest of the country. This newly available oil was refined in the United States into petroleum products, such as gasoline and diesel fuel, and exports of these products continued to rise.

In addition, new oil supplies have been identified and exploited in other countries. China, Russia, and several Middle Eastern countries all increased their production in the past decade through the spread and cost advances of such technologies as hydraulic fracturing and enhanced oil recovery. At the same time, while the economies of China and India continued to grow, the extremely high growth rates of the 1990s and 2000s were replaced with more-modest rates, but both countries also expanded energy-efficiency regulations. As a result, energy demand from the developing world has been growing more slowly than previously.

Sustained natural gas production from shale resources fueled large gas pipeline expansion projects from gas-producing regions to demand centers, as well as newly opened natural gas export facilities along the Gulf Coast. Exports of U.S. liquefied natural gas, fueled by demand in Asia, kept natural gas prices from falling to a level at which widespread use of natural gas for passenger or truck transportation would be adopted.

Although gasoline prices had begun to soar not long before the Great Recession, they bounced around for a few years, and, once the Alaska oil began flowing in 2016, they went from inching down to decreasing at a fast clip. By 2017, the average price per gallon was about $3.25—noticeably lower than the 2012 price of around $3.75 to $4, depending on the region—and the huge bite it had been taking out of drivers' wallets eased to the point at which it ceased to be a topic of conversation, either around kitchen tables or in policy discussions.

With prices getting lower, consumption naturally rose, albeit not dramatically. In 2009, Americans were consuming about 19 million barrels per day. In the ensuing two decades, this rose to 20.6 million barrels. In addition to the lower prices, other factors contributed as well. CAFE standards increased steadily through 2025, so the total vehicle fleet became more efficient. Also, lower oil prices kept airline ticket prices stable, encouraging demand for air travel. For trucks, fuel-efficiency standards did not increase nearly as much so consumption in the freight sector grew. (In general, freight movements are up, having increased 1.1 percent annually.) Other industrial processes adopted higher-efficiency practices as well.

Partly as a result of low oil prices, around the late 2010s, the limping economy got back on its feet. Unemployment finally began declining to a point that most Americans found tolerable, so that, by the 2016 presidential election, neither high gasoline prices nor high unemployment was much on the agenda. A key factor was that demand picked up as China successfully convinced its rising middle class to start thinking of products previously deemed luxuries as necessities—a category that included cars, as well as high-end foods and toiletries—opening up additional markets for U.S. businesses. Europe also successfully navigated the euro-zone crisis, so, by 2017, the world economy finally pulled out of the Great Recession entirely.

The American economy has been growing reasonably well, 2.5 percent annually, ever since. Service jobs catering to the affluent young elderly have boomed. The healthier among this group spend freely on travel and "beauticeuticals," products to make them look and feel younger. Those with more-serious medical issues spend on health aides, many of them immigrants, so they can continue to live at home. Given these demands, service employment is approaching 60 percent of all jobs, continuing its long-term increase from just over 50 percent in 2010. Immigrants began coming in steady numbers once Congress passed immigration reform. They continue to be willing to take the jobs that Americans might not want, although, given the affluent aging population, such jobs more often involve changing bedpans than picking lettuce.

Cheap Driving Induces Even More Suburbanization

Other trends historically associated with cheap oil have returned as well, such as suburbanization. The incredibly low interest rates introduced during the Great Recession continued for longer than they were probably needed, making it cheaper to buy houses. As people make more money and there are fewer long-term unemployed, new house sizes have increased, and people are moving to developing areas where land is cheap enough to allow larger houses and lot sizes. As gasoline prices have stabilized, driving is up. The minor shift toward city living that young adults led in the 2000s and 2010s began abating as they started getting married and having children, trading in their lofts and Zipcar memberships for freestanding houses and car ownership.

Cities have not been doing badly, population-wise, especially as more immigrants, drawn by the growing service economy, have arrived from countries with much-higher densities, where an 800-square-foot apartment would be an amazing luxury. Several Rust Belt cities—where heavy industries, such as steel, had long moved on—with previously declining populations reinvented themselves by attracting immigrants, who started up small businesses and had enough children to repopulate the public schools.

Climate Change Is Happening, but Effects in the United States Are Localized

The effects of climate change began to manifest themselves in the 2010s, but the size and diversity of the country meant that few things affected most Americans. Major storms, such as Superstorm Sandy in fall 2012 and Hurricane Hermione in 2016, had devastating effects on the regions where they made landfall, but it was easy enough for the rest of the country to tune out once the initial news coverage died down. The droughts of the early 2010s had some impact on food prices, but, once they returned to their predrought levels around 2015, the issue remained low on the national political agenda.

Environmentalists and scientists continued to sound the alarm, and they prodded some legislators to take action, but ultimately progress at the federal level has not occurred. Several climate bills were introduced in Congress, most notably the comprehensive Green Cap legislation of 2022, but also several with more-targeted emission reductions for specific sectors. None has cleared both the Senate and House, where a vocal minority of members continue to protect the coal and oil interests among their constituents. The gasoline tax was raised by $0.05 as part of debt-reduction legislation, but transportation officials saw this as too little too late. The Green Party, for years a very minor party whose only electoral victories were a few local officials, has grown to some extent, winning races in several state legislatures. However, with no major changes in the federal electoral system, it remains difficult for third parties to win seats at the national level.

To the extent that any policies to avert or mitigate climate change have been adopted, they have been taken by states and cities. As the country became more politically polarized, the population began sorting even more by political affiliation, meaning that the red states are even redder and the blue states bluer. Especially in coastal areas, some states have passed fairly stringent climate policies. Examples include stricter vehicle emission and fuel policies, incentives for household adoption of solar energy, and some limits on new energy producers, such as coal plants. But renewable energy sources remain at only 10 percent of all power generation—the same as in 2010. Policies remain somewhat fragmented, and generally only the most-committed households and cities are taking such actions.

Americans Are Driving More, and Many Roads Are Getting Worse

Something similar has happened in the transportation sector. At the federal level, the gasoline tax has been increased only once in the past two decades, and the amount of user revenues raised at all levels of government is still about $34 per 1,000 VMT, essentially the same as in 2008. This created the opening for states to play a larger role funding the roads within their borders. A few states have adopted modified mileage fee systems, especially where electric and natural gas vehicles were cutting into state gasoline tax returns. A few big cities have gone even further with various types of congestion pricing and higher tolls for "dirty" vehicles. But the majority of Americans live in places where few costs are attached to driving, with the exception that more toll roads and high-occupancy toll (HOT) lanes have been introduced along congested major corridors. The overall cost of driving remains unchanged since 2012, at about $0.52 per mile.

On average, expenditures on roadways (including federal, state, and local) are around $65 per 1,000 VMT, essentially unchanged since 2008. The results of these policies are most visible on a cross-country drive. Some states have great roads, frequently repaved and scenic. But cross a border into a neighboring state, and the disparity is obvious. Suddenly, traffic slows as drivers try to avoid serious potholes, as well as the truck operators who are doing the same. Rest areas are mostly closed, and a few states have detours around bridges that have been deemed unsafe.

When a state threatened to close a major bridge over the Mississippi River for safety reasons, the federal government stepped in to take it over and assume responsibility for building a replacement. This has happened only once, and it is not clear whether the federal government could consider this step for roads and bridges that are not part of the interstate system or federal-aid primary road networks. Along state roads, many small towns have emptied out because the long-distance travelers and truckers no longer stop in them.

The same thing has happened at airports. Those with high volumes of international travelers have been able to keep up, charging fees that fall mainly on nonresidents and becoming major retail hubs (helped, of course, by technologies that allow shoppers to try on clothes or play with toys in a realistic three-dimensional experience and that can ship the goods within a day). Smaller, regional airports have suffered the most, with the unluckiest having to close runways and reduce services. Air travel remains popular, with ticket prices remaining relatively stable because fuel is still reasonably cheap. However, the increasing air travel demand has caused capacity issues and frequent flight delays because airport infrastructure and air traffic control development have not kept up with passenger growth.

Congestion Has Increased, but Drivers Are Less Concerned

Not surprisingly, between suburbanization and cheap gas, congestion has grown. As noted above, relatively few Americans live in areas with road pricing. Because the road pricing policies have imposed fees that are politically acceptable—that is, affordable to most drivers—the experience has been that road pricing reduces rush-hour congestion modestly but not dramatically. In the rest of the country, revenues to fund some needed improvements remain scarce, so congestion is rising in most areas, just more in some than in others.

Years ago, drivers might have been up in arms about increased congestion, but technologies have made it easier to be productive while stuck in traffic. Smart-phone capabilities have evolved, and most smart phones now communicate seamlessly and easily with almost any post-2020 vehicle. Like many offices, vehicles operate with a bring-your-own-device (BYOD) ethos, under which almost any smart phone works in almost any vehicle, so drivers are not buying expensive and quickly outdated in-vehicle systems. But they can talk on the phone and send texts while driving, thanks to vastly improved voice-recognition software. In-vehicle cameras and windshield displays even allow telepresence, so drivers can conduct in-vehicle meetings.

Today, cars are much safer, even with all the new distractions. Safety sensors steer the vehicle back if it drifts into the next lane, and they warn the driver when the vehicle is following too closely. Adaptive cruise control helps the driver to keep a stable distance from the car in front of it. Like many safety features, such as air bags, these ADASs started in luxury models, but, by 2020, adaptive cruise control was standard on most new models, and other types followed within a few years. The distracted-driving problem, which caused a slight uptick in crash fatalities in the mid-2010s before these features became widespread, is less pronounced than it used to be.

The Vehicle Fleet Is More Efficient, Even Mostly Running on Gasoline

Another result of cheap gasoline is the relatively low percentage of AFVs. When gasoline prices edged past $4 per gallon in the 2000s—a major psychological barrier at the time—some predicted that Americans would start buying far more plug-in hybrids, fully electric vehicles, and perhaps even cars powered by fuel cells or natural gas. More such vehicles are on the roads in 2030 than two decades before, when the first plug-ins were just being introduced, but they are still fairly uncommon. As of 2030, only 1 percent or so of all new car sales are non-gasoline powered.

One major factor is that gasoline prices have remained low and steady for a long time. That dampened the immediate consumer demand not only for vehicles with greater fuel efficiency but also for the research that might have brought them down in price. Batteries remain heavy and expensive, and natural gas applications remain confined to specialty fleets, such as buses. The price difference from conventional vehicles remains high, limiting the demand to the same group of committed environmentalists who adopted the early Priuses.

AFVs simply have not gone mainstream, partly because of cost and partly because the available models are not particularly appealing to average car buyers. They remain an important niche, with the occasional prediction that this or that model will "break out," which has not happened yet. Drivers who own electric plug-in vehicles tend to recharge at home; a few competing firms install the upgraded outlets needed for at-home charging. The 18,000 or so public charging stations are clustered in cities where environmental awareness is high, generally at worksites or transit stations.

Although gasoline-powered vehicles are still prominent, mileage is much higher since the average CAFE standards of 54.5 mpg (4.3 L/100 km) were reached in 2025. The increase from 2012, when they were adopted, was so large that a few auto makers got out in front and introduced sleek new high-efficiency models that served the needs of families and older drivers, two of the main demographic groups buying new cars. Much of the fuel-economy improvements are due to lighter but stronger materials, and their cost has not increased the cost of vehicles by more than a few thousand dollars. So people are definitely paying more, but most are also keeping their vehicles longer because the average vehicle is generally good for 200,000 miles.

Transit Has Grown but Remains a Niche Market

Like alternative-fuel cars, transit occupies an important niche. Some cities that have attracted large numbers of immigrants have seen increases in ridership, along with those that have retained a vibrant core and neighborhoods. A few fast-growing Sunbelt cities completed new rail lines in the 2020s, and others have added a line or two to existing rail systems, but for the most part new riders have been on buses. Bus rapid transit has experienced a boom, favored by regions where the cost of new rail seems daunting and the right-of-way is available to add dedicated lanes. But many systems in poorer areas have accommodated new riders by adding new buses without replacing any existing rolling stock, leading to an older fleet on average and the impression of a system whose quality was slipping. Ridership is up in absolute terms, but only by 10 percent or so over the past decade.

Opportunities and Challenges Ahead?

The world has been a fairly comfortable place for Americans in the 2010s and 2020s. Low energy costs helped the economy shake off the Great Recession, more drivers are on the roads more safely and productively, and the population is growing (unlike in many other developed countries, where it is shrinking). However, the United States has not passed regulations to help address climate change and has missed an opportunity to use technological innovation to decouple the use of fossil fuels from economic growth. With major transport infrastructure deteriorating because of a lack of funding and with the consumption of fossil fuels still steadily growing, it is not clear whether robust economic growth will be sufficient to alleviate potential problems.

Chapter Four
Consequences for Future Mobility

The paths of mobility development illustrated by the two scenarios lead to alternative travel behavior outcomes. For each scenario, we developed estimates of PMT in 2030 for four transportation modes: vehicle, transit, domestic air, and intercity rail. These projections reflect the combined influence of multiple descriptors in the scenario, some of which tend to increase the amount of travel and others that tend to depress it. Our analysis takes into account the strength of these factors, as well as the size of their influence on travel. Details of how these estimates were developed are found in Appendix A. Using these PMT estimates, we also calculated the mode share for each of the four modes in 2030. Mode shares for 2010 are based on BTS (undated [b], Table 1-40). Results are shown in Tables 4.1 and 4.2. Although total travel increases in both scenarios because of population growth (both scenarios are based on the same population assumptions), the size of the increases varies by scenario and mode.

Table 4.1. Comparison of Total Passenger-Miles Traveled (in millions) at Baseline and in Two Scenarios

Transport Mode	Baseline (2010)	No Free Lunch (2030)		Fueled and Freewheeling (2030)	
	PMT	Change (%)	PMT	Change (%)	PMT
Vehicle, total	4,244,157	1.8	4,318,654	15.5	4,901,005
Transit, total	52,627	30.3	68,556	16.5	61,322
Domestic air	564,790	36.8	772,406	68.3	950,345
Intercity rail	6,420	17.6	7,551	9.4	7,024
All modes	4,867,994	6.1	5,167,167	21.6	5,919,696

Table 4.2. Comparison of Total Mode Share (percentage) at Baseline and in Two Scenarios

Transport Mode	Baseline (2010)	No Free Lunch (2030) Change		Fueled and Freewheeling (2030) Change	
	Share (%)	Direction	Percentage	Direction	Percentage
Vehicle, total	87.2	⬇	83.6	⬇	82.8
Transit, total	1.1	⬆	1.3	⬇	1.0
Domestic air	11.6	⬆	14.9	⬆	16.1
Intercity rail	0.1	=	0.1	=	0.1

NOTE: ⬇ indicates a decrease; ⬆ indicates an increase; and = indicates no change.

The differences in mobility outcomes between the two scenarios can be seen in the rates of growth for different modes in terms of PMT, rather than in changes in mode share. In the Fueled and Freewheeling scenario, total PMT has increased to 5.92 trillion, a 22-percent increase over the 2010 baseline of 4.87 trillion. The notable change among mode shares is the shift from highway to air travel; highway mode share has declined from 87 to 83 percent, while air travel share has increased from 12 to 16 percent. Transit and intercity shares remained steady.

In the No Free Lunch scenario, total PMT has increased by only 6 percent, to 5.17 trillion. Although all modes increase in PMT, the highway increase is far more modest than in Fueled and Freewheeling, from 4.24 trillion to 4.32 trillion. But in terms of mode shares, the differences between the two scenarios is slight. The decrease in highway mode share and the increase in the share of air travel are only marginally less pronounced than in Fueled and Freewheeling. There is a slight uptick in the transit share, while intercity rail remains steady.

To further explore the differences between the two scenarios and the factors contributing to them, we calculated the changes in per capita travel as well as in total. Figures 4.1 and 4.2 and Tables 4.1, 4.2, and 4.3 show the relative changes in total and per capita PMT, as well as changes in highway, transit, and air travel. Because intercity rail accounts for such a low mode share, we omitted it from this part of the analysis. Note that Figures 4.1 and 4.2 present PMT projections in index form, in which 2010 statistics for each mode represent a score of 100. Estimates greater than 100 indicate increases relative to 2010, while decreases are indicated by estimates lower than 100.

No Free Lunch: Total and Per Capita Mobility Growth Rates

In the No Free Lunch scenario, as shown in Figure 4.1, per capita highway travel has declined to 87 percent of its 2010 level. In addition to the increased share of the population over age 65, other factors contributed to this decline. Travel distances are shorter because of higher densities in both urban and suburban areas. Car ownership is lower, and a population with fewer vehicles tends to drive less. It is more expensive to drive: Oil prices and hence gas prices are up substantially, road pricing systems are in place in some metropolitan areas, and the GHG-emission policy is in place. Finally, far more employees take advantage of telework opportunities than did in 2010.

Transit use per capita has increased by 11 percent (see Table 4.3). A key reason for this is the increased density of urban and suburban areas, along with a steady increase in both the quality and amount of transit service. Coupled with the higher cost of driving and changing attitudes toward environmental protection, transit use in 2030 is considerably higher than it was in 2010.

Per capita air travel has increased by 17 percent, a substantial increase but far less than in the Fueled and Freewheeling scenario. Competing influences are at work here. On the one hand, economic growth has pushed air transport demand up. On the other hand, high oil prices in combination with additional CO_2 emission trading costs have increased ticket prices. The result is moderate growth in air travel.

This scenario assumes that some movement toward expanding intercity rail services has taken place by 2030. Given the increasing cost of air travel, as well as growing levels of airport congestion, several groups of states put renewed emphasis on developing high-speed rail. A consortium of midwestern states proposed a network of lines linking some key cities, and two of these lines have been constructed by 2030. Some rail links along the northeast corridor, which already had higher-speed Acela Express service in 2010, were upgraded to even higher speeds, and the first high-speed rail line opened on the West Coast. Although these developments induced some growth of intercity-rail PMT, its mode share remained at a very low level of 0.1 percent, the same as in 2012.

Table 4.3. Growth Rates for Transport Modes, No Free Lunch Scenario (%)

Transport Mode	Absolute		Per Capita	
	2010-2030	2010-2030, per Year	2010-2030	2010-2030, per Year
All modes	6.1	0.3	-9.5	-0.5
Vehicle	1.8	0.1	-13.2	-0.7
Transit	30.3	1.3	11.1	0.5
Domestic air	36.8	1.6	16.6	0.8

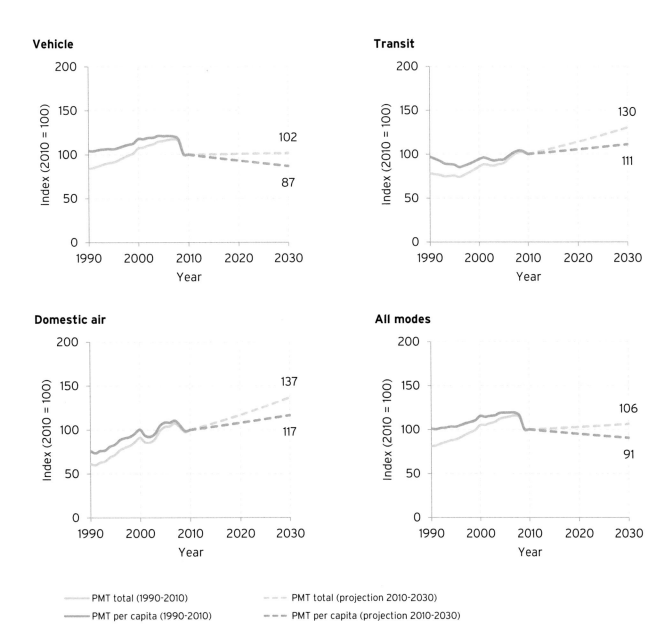

Figure 4.1. Projections for Passenger-Miles Traveled, No Free Lunch Scenario

SOURCES: For 1990-2010, BTS, undated (b), Table 1-40; U.S. Census Bureau.

NOTE: PMT projections are in index form, in which 2010 statistics for each mode represent a score of 100. Estimates greater than 100 indicate increases relative to 2010, while decreases are indicated by estimates lower than 100.

Fueled and Freewheeling: Total and Per Capita Mobility Growth Rates

In the Fueled and Freewheeling scenario (see Figure 4.2 and Table 4.4), per capita highway travel is projected to decline slightly from 2010 to 2030, even as total highway travel increases. One factor contributing to the decline of per capita highway travel is the aging population because older drivers tend to drive fewer miles than younger ones because the older drivers are not generally working. A second is increased congestion because more people are driving and the roads are in poor condition. Roads and bridges are in poorer condition in 2030 because of the lack of investment in maintenance and expansion, due to failures to increase taxes or mainstream road pricing. Finally, in this scenario, more people are working from home and shopping online than in 2010. But total vehicle travel is up because of the economy growing at 2.5 percent annually, higher rates of personal vehicle ownership, and only modest increases in fuel prices.

Transit use sees a similar pattern: slight decline in per capita use but an increase in overall ridership. More people own cars in this scenario, which depresses ridership slightly, but, on the other hand, some people in areas with severe congestion switch to transit.

Finally, after a five-year period of slowed growth due to the aftermath of the financial crisis, air travel has continued to grow substantially from 2015, on both per capita and total bases. This is due to both strong economic growth and moderate oil prices. Carriers also continued to drive down costs by better matching flight destinations, flight frequencies, and aircraft type with demand. Consequently, airfares have grown more slowly than inflation, which drove demand. Increasing air travel demand has caused capacity issues due to lacking airport and air traffic control infrastructure developments that slightly constrained air transport growth.

Table 4.4. Growth Rates for Transport Modes, Fueled and Freewheeling Scenario (%)

Transport Mode	Absolute		Per Capita	
	2010-2030	2010-2030, per Year	2010-2030	2010-2030, per Year
All modes	21.6	0.9	3.7	0.2
Vehicle	15.5	0.7	-1.5	-0.1
Transit	16.5	0.8	-0.6	0.0
Domestic air	68.3	2.6	43.5	1.8

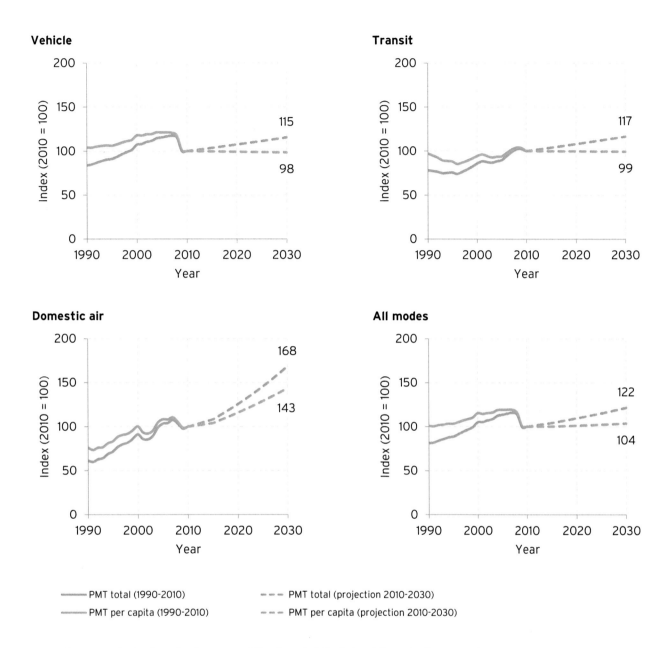

Figure 4.2. Projections for Passenger-Miles Traveled, Fueled and Freewheeling Scenario

SOURCES: For 1990-2010, BTS, undated (b), Table 1-40; U.S. Census Bureau.

NOTE: PMT projections are in index form, in which 2010 statistics for each mode represent a score of 100. Estimates greater than 100 indicate increases relative to 2010, while decreases are indicated by estimates lower than 100.

Driving Forces in the Scenarios

The two scenarios outline different possible paths to explain how mobility might develop in the United States on a national level up to 2030. Similarities exist in these paths, and these similarities represent developments with high probabilities because they resulted from projections on which the experts had inordinate consensus. For example, in demographics, under both scenarios, total population has grown at 0.8 percent annually since 2010 to reach 360 million people in 2030. The population has also grown noticeably grayer and browner—that is, both older and with a smaller share of white residents. In terms of economic developments, both scenarios indicate that overall GDP growth has strengthened, with annual growth rates between 2.0 and 2.5 percent. Income inequality has also increased in both. From a regulatory perspective, new CAFE standards, covering a period up through 2025, were passed in 2012. Similarities are also found in the technology field, in which advanced auto-motive technologies, such as vehicle-device interfaces and driver-assistance systems, became fairly widespread. However, both scenarios assume that fully automated driving has not become widespread by 2030.

Although similarities between the two scenarios exist, what is important for anticipating and preparing for change are the critical uncertainties, or driving forces, that cause one path to emerge over another (see Table 4.5). To identify these, we began with the information in Figure A.2 (see Appendix A) showing how active or passive each descriptor is. An active descriptor influences many other descriptors, while a passive descriptor is influenced by many others. The most active descriptor is economic growth, but both of our scenarios assume fairly high growth rates. However, two of the other highly active descriptors varied between scenarios: the price of oil and the presence of GHG regulations. The price of oil has been both high and volatile for several decades, so either of these prices is plausible, while the presence of GHG regulations is a political question.

Our third factor, the amount of highway revenues and expenditures (which consists of two descriptors), was not highly active according to this analysis. However, in reviewing the clusters of projections, we determined that highly passive outcomes, such as congestion and oil consumption, would be particularly influenced by revenues and expenditures. In addition, this uncertainty is less a continuation of a past trend (like oil prices) and only in part a political question (in the sense that it depends in part on the willingness to raise taxes). Instead, we seem to be at an inflection point in this area. Long-term trends in transportation revenues and expenditures are currently in flux (as discussed in Appendix F; see Brownell et al., 2013), meaning that a wide range of outcomes is possible. The future development of these three critical uncertainties will strongly affect other descriptors with which we dealt in our scenarios.

Table 4.5. Driving Forces with High Uncertainty in Future Development

Driving Forces	No Free Lunch	Fueled and Freewheeling
Oil price	High ($190/barrel)	Low ($90/barrel)
Level of environmental regulation	High	Low
Amount of highway revenues and expenditures	High	Low

The most speculative factor is the future price of oil. The market-based oil price results from the interplay between supply and demand and depends on various determinants, such as the economic development of industrial countries, geologically limited oil production, economic efficiency of (unconventional) oil production, amount of output from refineries, and the existence of political conflicts or incidence of natural catastrophes. Although the oil price in the Fueled and Freewheeling scenario remains on a volatile but low level, somewhere in the range of 2012 prices, the No Free Lunch scenario features a doubling of prices through 2030.

In both scenarios, domestic oil production has increased quite substantially to 15 million barrels per day in 2030. In combination with ongoing strong demand from rapidly developing countries, other factors led to the high price differential between the two scenarios. As a first difference, the No Free Lunch scenario assumes that new (unconventional) production technologies that many observers thought would bring down the price of drilling did not develop as hoped. Additionally, environmental regulation, such as the introduction of a national GHG emission-reduction policy, made fossil energy production more expensive. Furthermore, sustained and significant instability in the Middle East has meant supply disruptions, leading to price spikes as producers scrambled to keep up with demand. By contrast, the Fueled and Freewheeling scenario assumes that new production technologies have fulfilled their promise and brought down the cost. Here, the substantial increase in local U.S. oil production enabled oil prices to stabilize on a moderate level even lower than in 2012, despite ongoing global demand, especially from China and India. In all, the price of oil had a strong impact on U.S. oil demand in both scenarios, which differs between 16 million and 20.6 million barrels per day in 2030.

The second key uncertainty is the development of environmental regulation—in particular, the introduction of a GHG-reduction policy at the national level. Stricter regulation and the implementation of a national ETS in No Free Lunch have led to several changes in the energy and transportation markets. The subsequent doubling of the total share of nonhydro renewables up to 20 percent, combined with modified CAFE, fuel, and clean-energy standards, served as the basis for a considerable decrease of U.S. total CO_2 emissions through 2030. These changes in environmental regulation have been possible because of shifts in the recognition of climate change and its consequences on daily life for the American public. Although climate change has also been happening in the Fueled and Freewheeling scenario, its effects have manifested only in certain regions. In this scenario, climate-change issues have rarely surfaced on the national political agenda, although some states introduced fragmented emission-reduction policies on the state or city level. Overall, energy prices remained low because pressure for changing course on energy policy was not high enough to spur legislative action.

The third major uncertainty is highway revenues and expenditures. Sustained infrastructure investment is an indispensable requirement for an efficient transportation system. In Fueled and Freewheeling, highway revenues have stayed essentially the same as in 2008. The gasoline tax has been increased only once through 2030, and additional highway revenues have been mostly limited to local attempts to introduce various types of congestion pricing and higher tolls for dirty vehicles. The gap in sustainable infrastructure funding has widened between 2010 and 2030, and it has not been sufficiently filled by states and local entities. Driving in this scenario is relatively inexpensive, and infrastructure investment shortfalls have resulted in worsening of roads and rising congestion issues. No Free Lunch presents a different picture, in which road pricing has gone mainstream, providing essential new highway revenues. This outcome came about as elected officials and transportation planners observed that congestion pricing both raised revenues and helped to manage system performance, and they made stronger and more-public cases for adopting road pricing. As the public began seeing reductions in congestion, along with better-maintained roads, its resistance to pricing lessened.

Table 4.6 summarizes all scenario descriptors and projections for both scenarios that were developed at the workshops.

Table 4.6. Comparison of Projections in the Two Scenarios

DESCRIPTOR	NO FREE LUNCH	FUELED AND FREEWHEELING
Demography		
Total population	360 million (annual growth of 0.8%)	360 million (annual growth of 0.8%)
Share of population by race/ethnic group	Growing share of Hispanics and Asians, share of whites has continued to decline	Growing share of Hispanics and Asians, share of whites has continued to decline
Age structure	The share of the population over age 65 has increased steadily	The share of the population over age 65 has increased steadily
Population density	Urban/suburban densification with stable population shares	Suburban densities stable as edges of urbanization are pushed outward
Vehicles per 1,000 population	Decreased	Increased
Average household size	Average household size has remained stable, but share of households with children decreased slightly	Average household size has remained stable, but share of households with children decreased slightly
Economy		
Economic growth	Dampened average annual growth of 2.0%	Continued average annual growth of 2.5%
Income distribution	Income inequality has continued to increase	Income inequality has continued to increase
Labor-force participation	Women's labor-force participation kept growing slightly, while men's rate decreased slightly	Women's labor-force participation kept growing slightly, while men's rate decreased slightly
Sector employment	Manufacturing continued to decline slightly but stabilized at 7–8%; service employment increased to 55–60%	Manufacturing continued to decline slightly but stabilized at 7–8%; service employment increased to 55–60%
Freight movement	Ton-miles increased annually by 0.9%	Ton-miles increased annually by 1.1%

Table 4.6. Comparison of Projections in the Two Scenarios—Continued

DESCRIPTOR	NO FREE LUNCH	FUELED AND FREEWHEELING
Energy		
Introduction of GHG emission-reduction systems	National GHG policy was introduced by 2022	No new GHG legislation has been adopted by 2030
Electricity power generation sources	20% share of nonhydro renewables	10% share of nonhydro renewables
EV-charging infrastructure	80–85% of EV charging is done at home; about 100,000 publicly available charging stations	90–95% of EV charging is done at home; about 18,000 publicly available charging stations
Electricity prices	The average real electricity price for all sectors is $0.18/kWh	The average real electricity price for all sectors is $0.13/kWh
Adoption of AFVs	About 40% of all light-duty vehicles sold in 2030	About 1% of all light-duty vehicles sold in 2030
Oil consumption	16 million barrels/day	20.6 million barrels/day
U.S. oil production	15 million barrels/day	15 million barrels/day
Oil price	$190/barrel in 2030 (Brent crude)	$90/barrel in 2030 (Brent crude)
Transportation funding and supply		
Cost to drive per mile	$1.04 (doubling)	$0.52 (unchanged)
Mainstreaming of road pricing to increase revenue	In addition to priced lanes/facilities, variable parking pricing and MBUF systems are used in some areas	Only priced lanes/facilities are widely used
User revenues raised per mile driven	$45 per 1,000 VMT (30% increase)	$34 per 1,000 VMT (essentially unchanged)
Expenditures on roadways per mile driven	$80 per 1,000 VMT (30% increase)	$65 per 1,000 VMT (generally stable)
Congestion	Has increased only slightly	Has increased significantly
Quality and quantity of public transit	Transit service (measured in revenue miles) has increased by a total of 35%, and quality has increased[a]	Transit service (measured in revenue miles) has increased by a total of 10%, and quality has decreased[a]

Table 4.6. Comparison of Projections in the Two Scenarios—Continued

DESCRIPTOR	NO FREE LUNCH	FUELED AND FREEWHEELING
Technology		
Market penetration for broadband	95% of households use broadband technology	95% of households use broadband technology
Telecommuting share	40% of workers telecommute	15% of workers telecommute
Online shopping share of retail sales	30% of retail purchases (by number of transactions) are made online	30% of retail purchases (by number of transactions) are made online
Development of data privacy regulations	Weak regulation: Data privacy regulations allow the collection of in-vehicle data used (e.g., in MBUF systems or PAYD insurance)	Strict regulation: Data privacy regulations do not allow the collection of in-vehicle data used (e.g., in MBUF systems or PAYD insurance)
Adoption of telematic services	95% of all new vehicles in 2030	95% of all new vehicles in 2030
Market penetration of ADASs	90% of all new vehicles in 2030	90% of all new vehicles in 2030
Market penetration of autonomous vehicles	Very low (no more than 5% share in total car fleet)	Very low (no more than 5% share in total car fleet)

NOTE: All prices in 2012 dollars. MBUF = mileage-based user fee. PAYD = pay as you drive.

[a] The figures reported here for quality and amount of public transit do not match those estimated in Tables 4.3 and 4.4 because those projections were based on the estimation methodology described earlier in this chapter, while the projections in the table were elicited at the expert workshops.

Scenarios can be bounded by what is plausible, believable, or imaginable today in order to form a cohesive story about the future. In our study, we limited ourselves to the development of two scenarios, No Free Lunch and Fueled and Freewheeling, to focus on the key differences in projections. But, in thinking about the future of mobility, we do not want to miss any discontinuities that, in retrospect, may emerge as more important. So, in this chapter, we present two wild-card or low-probability scenarios. Wild cards are designed to provoke thinking about the unthinkable. These assume that certain events have broken with otherwise-foreseeable trends to move the world in an unanticipated direction. The underlying assumptions of these wild cards originated from comments made at the five expert workshops, in which we asked the experts what events might confound the projections they had just made, as well as from internal discussions at the sixth workshop among RAND, ifmo, and outside experts.

One wild card is based on the possibility that China experiences a major debt crisis and ensuing economic stagnation, with economic and demographic impacts that profoundly affect the United States. The other assumes that autonomous vehicles, currently unavailable commercially and thought by our experts to be several decades away, experience cost reductions that make them marketable much sooner than anticipated, with attendant impacts on transportation. Both wild cards use the standard convention of presenting them from the vantage point of 2030.

Red Dusk: China Stumbles

Overview

The U.S. economy has been stagnant for a decade, following a major debt crisis and ensuing economic stagnation in the People's Republic of China. In the past four decades, China has become the world's second-largest economy, but its massive investments in infrastructure backfired when it became clear that they were funded with unsustainable levels of local government debt. In the wake of a wave of defaults, the Chinese government tightened banking regulations considerably, and Chinese growth sputtered as Chinese banks and companies focused on repairing their balance sheets in a classic case of a "balance-sheet recession." Subsequently, Chinese growth remained in the low single digits for nearly two decades as the Chinese economy struggled to rebalance. As a result, the world economy faltered because neither China nor the United States could serve as the engine of growth.

Downward Spiral in China

China's economy began to falter in the late 2010s, when a wave of local government defaults began. The roots of this financial crisis stretched back to the 2000s, when local governments started pursuing ambitious building programs. These included not only residential construction but also major infrastructure projects, such as roads, airports, bridges, and dams. The sheer size of these building projects meant that construction spending eclipsed foreign trade as the largest component of China's economy, and, by 2011, China was investing a substantial percentage of GDP into infrastructure, far exceeding what any other country spent (Barboza, 2011).

Several reasons explained this rush to build. First, local governments lacked other sources of funding, so much of their economic development depended directly on real estate development (property taxes, a major source of local government funding in many other countries, were introduced only in 2011 in a few major cities). Many inland cities—which had not enjoyed the extraordinary growth of the coastal cities during the 1990s—also saw modern infrastructure development as a way to compete for investments and workers. In addition, the promotion system in the Chinese bureaucracy created incentives for local officials to pursue short-term economic objectives without adequate attention to long-term risks.

Local governments financed this building boom through borrowing. Although Beijing had long tried to rein in municipal borrowing through a prohibition on bond issuance, localities found ways to circumvent these restrictions by setting up off-budget investment companies that nonetheless acted as de facto development arms of the local government. By 2010, local governments officially had $2.2 trillion in combined indebtedness, though some observers believed that the true figure was closer to $3 trillion. These figures amounted to some 40-50 percent of the Chinese GDP (Barboza, 2011).

When the global financial crisis hit in 2008, China's stimulus package brought about yet more infrastructure spending. Credit became readily available through China's "big four" state-run banks, and the banking sector doubled in the ensuing three years. The building projects, already at a breakneck pace of development, added to the incentive to overinvest in infrastructure, resulting in areas of overcapacity. Housing and land prices began to inflate, leading to fears of another real estate bubble, similar to the one that sparked the U.S. financial crisis (Toh, 2012).

Initially, however, fears of a financial crisis were largely dismissed because Chinese officials felt confident that the country's banking sector was well-insulated from the sort of international bank runs that afflicted Asia during the 1997 financial crisis. The Chinese renminbi remained nonconvertible, and the bulk of foreign direct investments in China took the form of factories on the ground rather than portfolio investments. China's enormous foreign currency reserves of more than $3 trillion also gave Chinese policymakers confidence that they could step in and nip any emerging crisis in the bud. But what happened instead was not the proverbial run on the banks or a single localized crisis but a wave of local government defaults that caught Chinese central bankers off guard.

Although local banking crises are not new and have been happening with increasing frequency, in the late 2010s, a series of local crises—originally triggered by the collapse of a leading regional property developer with extensive loan guarantees in several "second-tier" cities in south-central China—coincided with each other in a perfect storm of unprecedented scale. Provincial authorities rushed in to save lower-level governments from insolvency. Social disturbances dramatically increased in the affected areas as laid-off workers took to the streets to demand back wages and unemployment benefits. When several major state-owned enterprises (SOEs) in the region were threatened with bankruptcy, the central government decisively stepped in with massive bailouts.

Although the bailouts were successful in stabilizing the situation, Chinese central bankers were chastened by the extent of the crisis. To forestall another crisis, the People's Bank of China (PBC) dramatically tightened banking regulations, and the big four commercial banks were ordered to repair their balance sheets by identifying bad debt and curtailing lending. The tightening of credit resulted quickly in a sharp decline in real property prices, despite government efforts at achieving a soft landing. The bursting of the real estate bubble inevitably resulted in more bad loans being written off. The PBC subsequently purchased much of the bad debt from the big four through a program of quantitative easing while depreciating the value of the renminbi after years of steady increases.

Under the new reactionary credit regime, many firms saw their net equities fall below zero (i.e., the value of their assets, often held in the form of real properties, fell below the value of their liabilities) and began to divert business earnings to paying down debts rather than making new investments. Although Chinese officials had hoped that the lower value of the renminbi would help stimulate exports and thus make up for the reduction in the investment component of the GDP, Chinese exports were hampered by weak global demands and the protectionist trade policies adopted by the United States following a period of high unemployment. Growth in China fell to the low single digits, a far cry from the days when it routinely topped 10 percent annually. With companies reluctant to invest and global demands for Chinese exports falling, Chinese policymakers struggled to rebalance the economy toward domestic consumption, which was slow to pick up because of flagging consumer confidence in the wake of the property price collapse. The world had hoped that growth in Chinese consumer demand might help stimulate the global economy, but the opposite happened: Demand dried up, the size of the Chinese middle class shrunk as many left for countries with more opportunity, and China became mired in a period of low-single-digit growth that continues today. This plunged the world into an extended period of economic stagnation and instability.

Effects of the Chinese Crisis in the United States

The Chinese crisis spilled into the United States in several ways. China was the largest foreign holder of U.S. debt up through 2020. But because it struggled with its own internal crises, China was no longer able to purchase the high number of U.S. Treasury bonds that helped finance the growing debt in the United States. Following the debt crises and bailouts of several poorer European countries, the United States has had trouble finding another cheap source of foreign capital. Since 2025, the United States has been forced to raise interest rates repeatedly to raise money from foreign sources. With little money for stimulus and a tight domestic credit market, the U.S. economy has been in recession since 2024.

Since the Chinese economic slowdown began, supply-chain uncertainties have grown exponentially, causing financial losses for the extended web of interrelated companies around the world. Unemployment has increased in the United States across all sectors as financial strains have continued, and most firms remain leery of large-scale new investments. Some manufacturing has come back to the United States from China because of rising concerns about security and stability, as well as protectionist trade policies adopted in the United States, but the impact on U.S. employment has been minimal given depressed domestic and global demands.

Another effect of the Chinese and East Asian slowdown is the increased influx of immigration from these countries. Since 2008, more newcomers to the United States have been Asian than Hispanic. The share of the U.S. population made up of people of Asian ancestry has grown to about 12 percent (more than double from 2010, when it was less than 5 percent). The majority of these immigrants are young adults with relatively high levels of education, congregating in urban areas where there are clusters of family and friends. Although immigration from China has increased steadily since the early 2000s, the recession and the associated social instability have greatly accelerated the influx, creating a spike in population density in major urban centers, such as Los Angeles, New York, and San Francisco.

Transportation Impacts of the Chinese Crisis

All infrastructure spending, including that on transportation, has stagnated in the past decade. Public- and private-sector funding, as well as foreign investment, have been tight to nonexistent. With deficit-reduction priorities and no money for stimulus, the only recourse for states and local governments to fund transportation improvements has been road pricing. But the public's and politicians' appetite for new road pricing projects has been curtailed as existing projects have increased user fees over the past decade to cover maintenance and operating costs. Road conditions throughout the nation have seriously declined, even on some key interstate routes.

Vehicle ownership has gone down because of the recession, as has passenger VMT. With incomes stagnating and high unemployment, Americans are buying fewer cars, holding onto their cars longer, and driving fewer miles for both commuting and shopping trips. Because of this, overall congestion has been modestly reduced. But with less funding for general maintenance, nonrecurring congestion (that is, congestion due to crashes) has gotten worse because crashes due to poor road conditions (such as sinkholes) have increased, and many states have reduced the emergency services that would clear crashes quickly.

Public transit use has increased with the recession and influx of immigrants to urban areas. The proportion of choice riders (those who could otherwise drive) has decreased, and the share of captive riders has increased. With less money for maintenance, service quality has worsened, causing widespread system delays. People continue to ride public transit, but their antipathy toward it has deepened. For the past five years, transit riders have voiced the opinion that they will stop using transit as soon as they can afford to. The possibility of a mass exodus of ridership once the economy improves will further erode support and use of public transit for the future.

Freight VMT has also decreased with the recession and supply-chain disruptions. West Coast ports have encountered lower demand as imports from Asia have decreased substantially, and East Coast ports are exporting fewer goods. Diminished travel demand in the United States (and elsewhere globally) has reduced oil prices, and U.S. oil production has been lower for many years because China was once the second-largest consumer of oil after the United States.

Deficit reduction has colored all political decisions. GHG policy has taken a back seat to other priorities and necessities. The money to support research and investments in AFVs and the necessary infrastructure to support them has dried up, and few Americans can afford them in the middle of a recession. Given the strong interdependence between China and the United States, it is hardly surprising that the effects of the China debt crisis have been so far-reaching.

The Autonomous-Vehicle Revolution

Overview

Autonomous vehicles have entered the mainstream by 2030, much more quickly than predicted, with about 15 percent of the fleet being autonomous. The key reason is a technological breakthrough that greatly reduced the cost of sensors. They have been judged safe, are legal for on-road use in all states, and have provoked several key changes in transportation.

How Autonomous Vehicles Entered the Market

The first autonomous vehicle—the Google Car, now in the National Museum of American History—was licensed in 2012. Autonomous vehicles became available commercially in 2014, but the early models were prohibitively expensive (several times the price of high-end luxury vehicles) for any but the wealthiest car buyers. Costs were high because the sensor technologies were custom-developed for niche markets and required precision manufacturing. However, as sensor technology began spreading to other systems, more research dollars were devoted to developing less expensive versions. In 2016, an academic research team invented a solid-state phased-array sensor that worked just as well as earlier mechanically scanned lidar sensors but that could be manufactured at scale at a lower cost.

After this technological breakthrough, commercial versions became more affordable. Once one automaker put out an autonomous vehicle in 2017 that cut the previous price by two-thirds, others quickly got into the game. Early versions ran on gasoline, but, within a few years, as batteries came down in price, the majority operated either as plug-in electric or on natural gas.

It was still nearly a decade before autonomous vehicles were allowed in most states because of insurance and licensing regulations. Many states with heavily urbanized populations were reluctant to change their regulations, especially after several early and heavily publicized crashes caused by malfunctioning sensors. In the late 2010s, autonomous vehicles—or aut-Vs, as they are now commonly known—were legal only in the sparsely populated western states.

By 2022, however, more than half the states allowed aut-Vs on the roads, and the last holdout legalized them in 2026. Several groups were instrumental in pressuring states to legalize them. Advocacy groups representing the disabled and the elderly were the most vocal, pointing to the increased quality of life for people who would be able to travel independently, not to mention the cost savings of allowing people to live more easily at home. The auto manufacturers themselves pushed for them. Even transit agencies, which had concluded that some types of service could be provided more cheaply through aut-Vs than conventional transit, wanted to see them on the roads. As costs declined (the average aut-V was roughly 25 percent more expensive than a conventional car by 2026), the pressure to mainstream them increased. The crash rate had also declined, providing some political cover to nervous state legislators.

States have changed the legal environment sufficiently to allow aut-Vs to operate on all public roads. One compromise most states reached was that the vehicles needed to operate with at least one occupant. In a small number of states, policies require that aut-Vs could travel no more than a short distance with no occupant. Although no state requires a special license to use an aut-V, most states prohibit solo riders under a minimum age or those who have been declared mentally incompetent. This gave rise to a new type of minimum-wage job, the aut-V minder, who basically rides for hours at a time to populate the vehicle while it drives from one passenger drop-off to the next pickup. State legislatures and courts have also had to address liability laws. Lawsuits have been filed following serious crashes, alleging that the auto manufacturers were at fault; although some have been successful in winning damages, others have not. Many related areas of law remain unsettled, such as the privacy of data collected from drivers. Hacking remains a problem, although most cases have involved celebrities or estranged spouses; most people in those categories simply do not use aut-Vs.

Today, in 2030, about 15 percent of all vehicles are aut-Vs—enough to have changed, but not revolutionized, transportation. Some services have been transformed, especially in the cities, but the majority of Americans are still driving themselves and over similar distances.

Transportation Impacts of Autonomous Vehicles

A few key and fairly affluent groups were early adopters. First, parents bought them for their teenagers. Teenagers still needed to get to school and work, and stricter licensing standards and stiffer penalties for texting while driving meant that fewer teenagers were driving themselves. By 2028, the percentage of 19-year-olds without a license passed 50 percent, continuing a trend that started in the early 2010s. A skill that used to be nearly universal—the ability to drive—is fading among this generation.

The elderly made up another key group, seeing a chance to keep living the suburban dream even as their doctors revoked their driver's licenses. Aging baby boomers, now in their 70s and 80s, have been able to remain independent because loss of a driver's license no longer means relying on grown children or expensive taxis for rides. Finally, a third group, albeit the smallest, were the "super-commuters," people who did not mind driving 90 minutes or more each way when they could turn their vehicles into a true mobile office. Mobile devices allow them to carry work everywhere and seamlessly, from home to car to office with almost no disruption in e-services. Early predictions that aut-Vs would allow people to live and work anywhere proved to be exaggerated, not unlike the dream that computers would eliminate paper.

In response to these markets, automakers began radically changing vehicle designs. The newest mobile-office vehicles now come equipped with in-vehicle web access, telepresence connections, and high-resolution display screens. Vehicles serving the elderly and disabled have low-floor entry, a medical emergency call button, and some basic bio-scan features so tele-nurses can continue getting the medical feeds they might routinely monitor.

Aut-Vs also made new types of transportation services possible. First, car-sharing got a huge bump as it combined car-sharing with ride-sharing. The pricing structure makes them generally less expensive than taking taxis but more expensive than owning a vehicle. (Taxis went out of business fairly quickly when aut-Vs could beat them in price per mile; a few remain, but, over the protest of driver unions, most cities stopped issuing new taxi medallions by 2024.) The more affluent use private aut-Vs to commute, but, for many people, the main purpose is errands or going out in the evening, for which they use shared aut-Vs.

Second, transit agencies started using aut-Vs to provide service on low-use bus routes at lower labor and fuel costs. Transit agencies provide eight- and ten-seat aut-V jitney-type service to serve primarily immigrant neighborhoods where workers need access to low-wage jobs clustered in malls and assisted-living centers.

Some elements of travel have changed for the better. Aut-Vs can and do crash because of malfunctions or hacking, most typically pile-up rear-end crashes caused by an aut-V freezing up while driving. But overall, roads are much safer now. Crash rates had been declining in the early 2010s, due in part to the Great Recession, when fewer vehicles were on the road, and have continued to fall with more aut-Vs on the road. Last year, just over 10,000 Americans died in car crashes, down from 33,000 in 2012. Technologies such as crash-warning systems have reduced some crashes, and aut-Vs have drastically reduced drunk driving. But traditional (self-driving) cars remain on the roads in large numbers, and many persons now learn to drive from shady for-profit private schools since high schools dropped mandatory driver education in the late 2010s.

Other elements have not changed much. Rush hour remains because the majority of commuters still drive themselves. In some areas, traffic is even worse with aut-Vs, despite widespread virtual road train capabilities that help reduce congestion, because overall miles have increased. Detailed aut-V data (miles traveled are centrally collected by transit operators and companies that run shared-aut-V services) show that commutes have gotten longer, among both aut-V owners and those who share rides. Affluent families hire multiple aut-Vs to ferry their children to school while they head to work, and the elderly have not curtailed their driving.

The key advantages of aut-Vs are the drop in fatal crashes, the quality of life for the older generations, and the reduced emissions because most alt-Vs run on electricity or natural gas or are programmed to drive efficiently. The remaining challenges include the security of the associated communication networks, which leaves them vulnerable to accidental disruptions, as well as hacking. The prevalence of aut-Vs seems poised to grow, and, with it, something closer to a transportation revolution.

Chapter Six
Implications of the Scenarios

Each scenario captures a hypothetical context in which future transportation policy and planning might be conducted. The scenarios account for both the current state of affairs (because the projections were based on past trends) and the various forces that may be shaping the future state of affairs in 2030. These forces, which may be more or less likely and more or less desired, will have a combined effect on mobility outcomes in the future.

As noted previously, our scenarios are descriptive and not normative. We did not seek to define a desired mobility future and then identify the path to arrive at that future. Such thinking is left for the different users of the scenarios. Instead, our scenario approach explored possible future developments with past trends as a point of departure. In other words, we tried to answer "What if?" and not "How to?" questions:

• What if clear and consistent evidence of climate-change events shaped popular and legislative support for stringent GHG emission-reduction policies?
• What if GDP and oil consumption were no longer coupled?
• What if road pricing were to go mainstream?
• What if consumers experienced sustained low oil prices?
• What if the American economy started booming as it pulled out of the recession?

In this chapter, we present the implications of the scenarios for transportation policy and transportation planning.

Implications for Transportation Policy

Our study focused on long-term scenarios for passenger travel, which includes travel by car, domestic air, and intercity rail. Long-term scenarios in this topic area are multilayered and complex, being influenced by demographics, economics, energy, transportation funding and supply, and technology. How these forces play out over the next 20 years will depend on whether and how policymakers and other decisionmakers sort out and address current and upcoming challenges. Although we cannot know these outcomes in advance, we can apply scenario planning to develop plausible mobility futures that can be used to anticipate and prepare for change.

The two scenarios, No Free Lunch and Fueled and Freewheeling, illustrate the alternative futures that result when different policy directions are pursued related to economic growth, environmental regulation, infrastructure investments, road pricing, zoning and housing locations, transit maintenance and investment, new vehicle and energy technologies, telework, air industry consolidation, and congestion reduction.

In No Free Lunch, a confluence of policies and regulations related to growing concerns about climate change result in lower per capita highway travel due to shorter trip distances, lower levels of car ownership, and road pricing that creates a financial disincentive to drive. Transit use per capita has increased, along with a modest increase in per capita air travel. The No Free Lunch context has also nudged an expansion in intercity rail services.

In Fueled and Freewheeling, market influences have had greater influence than policy ones. Total highway travel has increased because the economy is booming, but per capita highway travel has declined slightly because of heavy congestion. Taxes have not been raised, and road pricing has not been implemented, in recognition of negative public and legislative sentiment regarding these two issues. But roads and bridges in some states are in poorer condition. There is a slight decline in per capita transit use because more people own cars, but an increase in overall ridership in areas with severe congestion. Air travel has grown substantially because of both strong economic growth and moderate oil prices.

Analyzing the differences between the two scenarios, we can identify the critical uncertainties, or driving forces, that caused one path to emerge over another. Our analysis revealed three factors as being significant in this regard: (1) the price of oil, (2) the development of environmental regulation, and (3) the amount of highway revenues and expenditures.

Price of Oil

Although the oil price in the Fueled and Freewheeling scenario remains on a volatile but low level, somewhere in the range of 2012 prices, the No Free Lunch scenario features a doubling of price by 2030. But the price of oil is exogenous. Transportation policymakers have virtually no leverage over it. The market-based oil price results from the interplay between supply and demand and depends on various determinants outside of transportation. The other two drivers are well within the purview of transportation policy.

Development of Environmental Regulation

Transportation decisionmakers and planners who value environmental sustainability have been promoting stricter environmental regulation—in particular, the introduction of a GHG-reduction policy on a national level. Because many believe that there is an economic cost to environmental regulations, policymakers need strong evidence of the benefits. The implementation of a national ETS in No Free Lunch has led to several changes in the energy and transportation markets. These changes in environmental regulation have been possible only because public attitudes have shifted to strongly support urgent action to avoid further negative consequences. Although climate change has also been happening in the Fueled and Freewheeling scenario, its effects have manifested only in certain regions. In this scenario, climate-change issues have rarely surfaced on the national political agenda, although some places have introduced fragmented emission-reduction policies at the state or city level.

Policies to Increase Highway Revenues and Expenditures

Policymakers have long been reluctant to support an increase in the gasoline tax or the implementation of other methods of raising transportation revenues. A core element in the sustainability of the nation's transportation infrastructure is the adequacy of highway revenues and expenditures. In Fueled and Freewheeling, highway revenues have stayed essentially the same as in 2008 because of an unwillingness to increase the gasoline (or other) taxes and to implement road pricing on a widespread scale. The result is that driving is relatively inexpensive (especially with low oil prices), which is good for consumers. However, congestion is worse, and some roads have not been adequately maintained. In No Free Lunch, road pricing and spin-off revenues from the carbon tax have increased highway expenditures. In addition, driving is expensive, and the amount of driving has decreased, so most states do not require as high a level of expenditures as in the Fueled and Freewheeling situation.

Implications for Transportation Planning

Our two scenarios present transportation policymakers, planners, transportation suppliers, and private-sector users of the system with the different challenges and opportunities that they might face under one scenario versus another. This information provides a context for understanding how today's decisions, among any of these players, might play out in the future. Here we suggest three ways in which to apply and use the scenarios in transportation planning. Depending on who is using the scenarios, different implications for planning can be drawn.

Identifying Early Warning Signs

One of the fundamental uses of scenarios is that, if considered plausible, they can help policymakers and other decisionmakers anticipate and prepare for change. The systematic, long-term view of different paths of mobility development supports creative but focused what-if thinking. As an initial step toward further action and planning, it is useful to develop a way to monitor key trends in relation to each scenario. Early warning signs (or leading indicators) of directions in which critical uncertainties might go can and should be discerned now and monitored over time. They can be categorized by the relative strength of their connection to demographic, economic, energy, transportation funding and supply, or technology issues. Considering all of the influencing areas when identifying early warning signs forces the acknowledgment of shifts in trends outside the transportation-specific domain. The purpose of this exercise is, then, to ask, "Toward which scenario are we moving, and what are the implications of this?"

Specific early warning signs can be developed on the basis of the key trends set out in the scenarios, supported by appropriate data sources that are monitored on a regular basis. For example, under the No Free Lunch scenario, potential early warning signs include the frequency of U.S.-based climate shocks, shifts in the national environmental agenda, momentum for carbon sequestration, and market demand for urban living. Under Fueled and Freewheeling, early warning signs are the cost of driving, strength of GDP growth, new home sales in suburbs, and air travel demand. Different users of the scenarios may be more interested in one category of early warning signs than another depending on their assumptions about critical uncertainties.

Determining Opportunities, Risks, and Contingencies

Because multiple scenarios force us to consider a wider range of futures than in typical day-to-day planning, scenarios serve to uncover new opportunities on the horizon and to highlight key risks. In this way, the scenarios presented in this report can be used to influence an agency (public or private) to consider a wider set of strategic options within its strategic planning process. Strategic planning typically begins with the desired end state and works backward to the current status. At every stage of strategic planning, the planner asks, "What must be done at the previous (lower) stage to reach this stage?" Making sense of past events and monitoring potential future developments when working in a pressured environment (as transportation often is) is a challenge. Scenarios enable strategic planners to look at a wider set of opportunities and risks, and therefore to identify a more robust set of strategic options.

One key reason is that scenarios are useful in acknowledging and representing systemic risks—that is, risks that are generated by a combination of factors. Traditional risk management tends to present and consider risks individually and can therefore miss these connected effects. For example, in Fueled and Freewheeling, the scenario captures the risks pertaining to future natural gas supplies in the United States and the vulnerabilities in terms of increased use as a vehicle fuel. According to the scenario, sustained natural gas production from shale resources fuels new gas pipeline expansion projects in demand centers, as well as in newly opened natural gas export facilities along the Gulf Coast, thus increasing supplies for export. At the same time, demand in Asia for U.S. liquefied natural gas has kept gas prices from falling to a level at which its widespread use could support adoption for passenger or truck transportation. Thus, in 2030, the vehicle fleet in the United States remains predominantly gasoline powered.

Our scenarios are, then, a useful platform on which to build contingency plans. These plans can be tested against the what-if projections embedded in the scenarios. Are contingencies robust and resilient over more than one scenario? If not, can they be adapted swiftly to cope with the challenges in both of the scenarios?

Reviewing Strategic Options Against Scenarios

By highlighting major challenges and risks as noted above, scenarios can provide a valuable reality check on current strategic options and plans. Then, a follow-up exercise can help focus a transportation policy or planning discussion on the critical uncertainties in moving forward toward a particular strategic vision. One way of reviewing options against scenarios is to map a set of strategic options against the scenario descriptors and projections in a simple matrix. The focus is on how robust each strategic option is (i.e., can it be delivered in a particular scenario?) and on its strategic importance (i.e., how important is it in influencing a particular scenario outcome?).

This method, sometimes called *wind tunneling*, is often used for testing the fitness of an idea or concept, much as a wind tunnel tests the fitness of an airplane or automobile design (van der Hiejden, 2005). Wind tunneling offers a process by which elements of a certain policy direction can be played against possible futures to reveal the different ways in which those elements might influence, and be influenced by, other factors in the scenario. For example, in the case of introducing a scheme of mileage fees in a particular state or group of states, several areas of technology and political and economic forces can be wind tunneled against various aspects of infra-structure investment needs. Where would the political support for such a scheme come from in No Free Lunch, and how would that be different in Fueled and Freewheeling? Who would implement the scheme in each of the scenarios? What would be the opportunity costs for the state in the scenarios? When the various aspects of such a scheme have been wind tunneled, states might develop a perspective on how those elements would fit together.

Utility of the Wild-Card Scenarios

Future planning, even long-term future planning, can be constrained by a too-narrow focus on what is imaginable today. Our scenario process, like many others, followed a systematic approach of drawing out and analyzing possible future projections on a specific set of descriptors based on past and current trends. The systematic approach provides greater credibility to the scenarios, but it does constrain the scenarios to what might be plausible given current conditions.

The value of the wild-card scenarios is that they escape the condition that they must be believable today—they are breaks in trends. They represent the extreme scenarios that totally redirect the paths identified for No Free Lunch or Fueled and Freewheeling. Although our study identified two such wild cards, Red Dusk: China Stumbles and The Autonomous-Vehicle Revolution, other wild cards should be considered in both transportation policy and planning, such as a triple-dip recession, a global pandemic, or a health technology breakthrough that produces extreme longevity.

Planners can use wild-card scenarios in the same ways they use the more-plausible scenarios. Planners can assume that those scenarios are possible, if unlikely, and plan accordingly. For example, in the transportation realm, autonomous vehicles are not in use commercially today and are not expected to be in the next two decades. But a state planning agency might develop a contingency plan for developing new insurance regulations around autonomous vehicles if it becomes apparent (for example, through sales figures of such vehicles) that widespread adoption will happen more quickly than anticipated. This would allow faster implementation of such regulations than in states where widespread adoption was assumed not to occur.

Chapter Seven
Conclusions

Our project sought to answer the question, "What might we expect for the future of mobility in the United States in 2030?" Knowing that the future of mobility in the United States in 2030 is uncertain, we developed two scenarios, No Free Lunch and Fueled and Freewheeling. These scenarios illustrate the paths that may result from interconnected impacts of market, policy, and consumer forces. No Free Lunch describes a future in which the United States has strengthened regulations to reduce dependency on oil and GHG emissions, which result in greater investment in AFV R&D, increased public transit ridership, greater reliance on road pricing, and lower levels of car ownership. Fueled and Freewheeling describes a future in which the economy is booming and a reluctance to raise taxes is prevalent, which result in high levels of car ownership and steadily increasing congestion.

We applied scenario planning, which is increasingly being used to deal with opportunities and risks of complex long-term issues, such as future mobility, instead of straight-line trend analysis or improved travel demand forecast models. Not only are the data to support these latter approaches incomplete and evolving, but also the accuracy of long-term forecasts has long been suspect. Predictions usually deteriorate with time because of unforeseen effects (Flyvbjerg, 2009). The relationship between today's situation and a long-term future outcome is hardly linear. It takes a systematic process, such as scenario planning, to identify possible, plausible futures and then explore the paths leading to those alternative futures.

As we look ahead to 2030, multiple mobility futures are possible. The paths of mobility development illustrated by the two scenarios lead to alternative travel behavior outcomes. The differences in mobility outcomes between the two scenarios can be seen in the rates of growth for different modes in terms of PMT. In the Fueled and Freewheeling scenario, the total number of PMT has increased to 5.92 trillion, a 22-percent increase over the 2010 baseline of 4.87 trillion. In the No Free Lunch scenario, the total number of PMT has increased by only 6 percent, to 5.17 trillion. In terms of mode shares, the difference between the two scenarios is slight. We found a decrease in highway mode share and an increase in the share of air travel in No Free Lunch that was only marginally less pronounced than in Fueled and Freewheeling.

The study identified three critical uncertainties, or driving forces, that caused one path to emerge over another: the price of oil, the development of environmental regulation, and the amount of highway revenues and expenditures. Of these, the most critical and speculative is oil price. Although the oil price in the Fueled and Freewheeling scenario remains on a volatile but low level, somewhere in the range of 2012 prices, the No Free Lunch scenario features a doubling of prices between now and 2030.

The second driving force is the development of environmental regulation—in particular, the introduction of a GHG-reduction policy at the national level. Stricter regulation and the implementation of a national ETS in No Free Lunch led to several changes in the energy and transportation markets. Although climate change has also been happening in the Fueled and Freewheeling scenario, its effects have been manifested only in certain regions. Some states introduced fragmented emission-reduction policies on the state or city level. Overall, energy prices remained low because pressure for changing course on energy policy was not high enough to spur legislative action.

The third major uncertainty is highway revenues and expenditures. In Fueled and Freewheeling, highway revenues have stayed essentially the same as in 2008. The gap for sustainable infrastructure funding has widened between 2010 and 2030, and it has not been sufficiently filled by states and local entities. Although, in this scenario, driving is relatively inexpensive, infrastructure investment shortfalls have resulted in worsening roads and rising congestion issues. No Free Lunch presents a different picture, in which road pricing has gone mainstream, providing essential new highway revenues. Mileage-based fees, priced lanes, and variable toll roads crisscross every state and metro area, with the result that the country has seen a 30-percent increase in highway revenues since 2008. As a result of these new revenues, the United States can collectively spend about 30 percent more on transportation infrastructure improvements. In Chapter Six, we point out that the potential for transportation policymakers and other decisionmakers to influence the price of oil is limited. However, they will have greater opportunity to leverage the other key drivers if there is public acceptance and political will.

What can we expect for the future of mobility in 2030? If we take the path of Fueled and Freewheeling, the future is a fairly comfortable place. We find low energy costs and a thriving economy. More drivers are on the roads more safely and productively. The population is growing, unlike in many other developed countries, where it is shrinking. But challenges loom on the horizon. On the No Free Lunch path, the United States is dealing with the issue of climate change directly, and, to the surprise of many, it has had a positive rather than negative effect on the economy. The United States has lowered its dependence on oil and, through road pricing, put the transportation system on a more stable financial footing. It has done both through the implementation of regulations, taxes, and fees, reflecting that there is always a cost to people or to society.

Our wild-card scenarios, which raise the specter of a major debt crisis and ensuing economic stagnation in China and bring the promise of fully automated driving to the market earlier than anticipated, point out that unexpected events—even those considered to be extreme outliers—could have major effects on the future of mobility. Assuming that such events are plausible and worthy of contingency planning, even though the probability of their happening is extremely low, is important for strategic policy and planning.

By making potential long-term mobility futures more vivid, our aim is to help planners and policymakers at different levels of government and in the private sector envision what the future might bring. In this way, they may better anticipate and prepare for change and, in the process, make better decisions now to affect the future of mobility in the United States.

This appendix describes in more detail the methodology used to develop the scenarios presented in this report. We define *scenario* as a plausible combination of possible future developments. Scenarios support what-if thinking and lay the foundation for alternative strategies to reduce uncertainties in mid- and long-term planning (Gausemeier, Fink, and Schlake, 1998; Lempert, Popper, and Bankes, 2003; Mietzner and Reger, 2005). Scenario planning is distinguished from forecasting in that it produces multiple potential futures, as illustrated in Figure A.1.

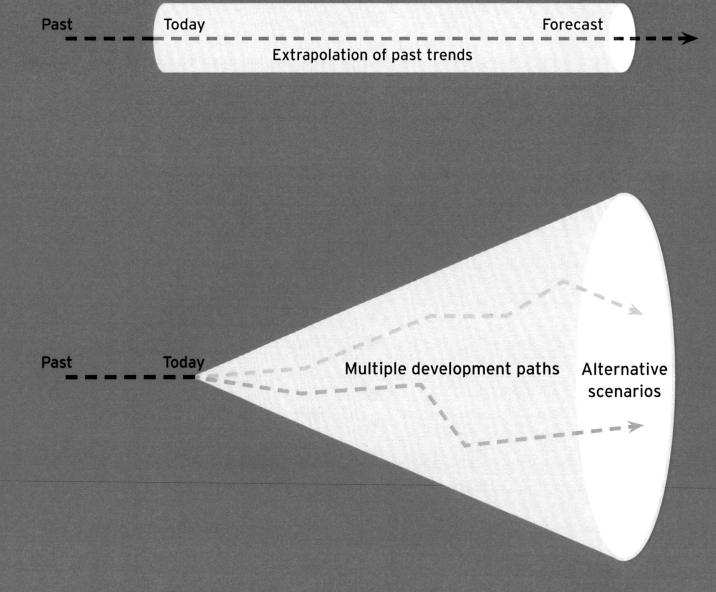

Past   Today   Forecast

Extrapolation of past trends

Past   Today   Multiple development paths   Alternative scenarios

Figure A.1. Differentiating Scenarios from Forecasts

SOURCE: Institute for Mobility Research.

Scenarios can be developed using several approaches (Mietzner and Reger, 2005). Early scenario techniques focused on the solely qualitative description of different futures (scenario writing) and used mostly intuitive approaches to arrive at these pictures and strategic statements (Kahn and Pepper, 1979). Over the years, different process steps were strengthened and formalized to address the complexity of strategic issues. Consistency analyses began to be used to measure scenario quality and relevance. Results were documented in narratives with statements relating to key indicators. Today, scenarios are developed with more-quantitative approaches that rely on multiple model runs and computer tools, which enhance the ability to cope with system complexity and make the resulting scenarios less arbitrary (see, for example, Gordon and Hayward, 1968; Kane, 1972; Gausemeier, Fink, and Schlake, 1998; Lempert, Popper, and Bankes, 2003; Bryant and Lempert, 2010; Rozenberg et al., 2012; Schweizer and Kriegler, 2012; Gerst, Wang, and Borsuk, 2013). Our methodological approach (outlined in the Introduction of this report and presented in more detail here) is representative of a more quantitative approach to scenario development.

Each of the six steps of the scenario approach is described in detail.

Step 1: Select Influencing Areas

In the first phase, the team identified influencing areas and descriptors relevant to building the scenarios. Influencing areas are topics germane to the scenario context. The team drew on previous ifmo research, as well as RAND research currently under way for the Transportation Research Board, to identify five influencing areas: demographics, economics, energy, transportation funding and supply, and technology. Next, we identified descriptors within each influencing area; these were also based on prior research of the study team. We define *descriptors* as indicators within an influencing area; they can be quantitative or qualitative.

For each influencing area, RAND experts produced a white paper, documenting past trends for each descriptor over a period of at least 20 years, or more if data were readily available. (A summary of past trends is presented in Chapter Two of this report, and the complete white papers are available as separate, web-only appendixes [Brownell et al., 2013].)

Step 2: Elicit Projections on Descriptors

Five workshops—one for each influencing area—were held in RAND's Washington-area office in April and June 2012. Participating in each workshop were six to eight outside experts, for a total of 37 individuals (see Appendix B for a list of these experts). In a facilitated discussion, we asked the experts to develop a projection for each descriptor in 2030. The projection could be qualitative or quantitative. Each expert estimated his or her upper- and lower-bound projection, followed by his or her best estimate. We asked them to provide reasons that a certain projection might be plausible and under what conditions. We also asked them to discuss any qualitative effect on travel behavior and mode choice.

Table A.1 shows all 32 descriptors and 61 projections developed during the five workshops. For some descriptors, the experts agreed on a single projection. For example, the demographic experts agreed that the U.S. population would grow at a particular average rate. For other descriptors, multiple projections were produced. In some cases, this was because opinions varied; in others, it was because the experts agreed that the future value of the descriptor would vary depending on other factors. For example, domestic oil production might vary with the costs of developing unconventional oil sources, as well as world prices and demand. The experts determined how many projections to produce for each descriptor.

Table A.1. Influencing Areas, Descriptors, and Projections

	DESCRIPTOR	PROJECTION
Influencing area: Demographics		
1.1	Total population	Moderate growth of 0.8% per annum
1.2	Share of population by race/ethnic group	Growing share of Hispanics and Asians; share of whites has continued to decline
1.3	Age structure	The share of the population over 65 has increased steadily
1.4	Population density	(a) Urban/suburban "densification" with stable population shares
		(b) Suburban densities stable as edges of urbanization pushed outward
1.5	Vehicles per 1,000 population	(a) Increased
		(b) Leveled off at 2009 rates
		(c) Decreased
1.6	Average household size	Average household size has remained stable, but share of households with children decreased slightly

Table A.1. Influencing Areas, Descriptors, and Projections—Continued

	DESCRIPTOR	PROJECTION
Influencing area: Economics		
2.1	Economic growth	(a) Slowed average annual growth between 1.5 and 2.0%
		(b) Continued average annual growth between 2.0 and 2.5%
2.2	Income distribution	Income inequality continued to increase
2.3	Labor-force participation	(a) Labor-force participation has remained generally stable, with men at about 70% and women at 60%
		(b) Women's labor-force participation kept growing slightly while men's rate decreased slightly
2.4	Sector employment	Manufacturing continued to decline slightly but stabilized at 7-8%; service employment increased to 55-60%
2.5	Freight movement	(a) Ton-miles increased annually by 0.6-0.8%
		(b) Ton-miles increased annually by 0.8-1.0%
Influencing area: Energy		
3.1	Introduction of GHG emission-reduction systems	(a) National GHG policy was introduced by 2022
		(b) No action had been taken in 2020, but policies were adopted by 2030
		(c) No new GHG legislation has been adopted by 2030
3.2	Electricity power generation sources	(a) 20% share of nonhydro renewables
		(b) 10% share of nonhydro renewables
3.3	EV-charging infrastructure	(a) 90-95% of EV charging is done at home; about 18,000 publicly available charging stations
		(b) 80-85% of EV charging is done at home; about 100,000 publicly available charging stations
3.4	Electricity prices	(a) The average real electricity price for all sectors is $0.13/kWh
		(b) The average real electricity price for all sectors is $0.18/kWh

Table A.1. Influencing Areas, Descriptors, and Projections—Continued

	DESCRIPTOR	PROJECTION
Influencing area: Energy		
3.5	Adoption of AFVs	(a) About 1% of all light-duty vehicles sold in 2030
		(b) About 8% of all light-duty vehicles sold in 2030
		(c) About 30–45% of all light-duty vehicles sold in 2030
3.6	Oil consumption	(a) 16 million barrels/day
		(b) 20.6 million barrels/day
3.7	U.S. oil production	(a) 15 million barrels/day
		(b) 6 million barrels/day
3.8	Oil price	(a) $90/barrel in 2030 (Brent crude in 2012 dollars)
		(b) $130/barrel in 2030 (Brent crude in 2012 dollars)
		(c) $190/barrel in 2030 (Brent crude in 2012 dollars)
Influencing area: Transportation funding and supply		
4.1	Cost to drive per mile	(a) $0.52 (unchanged)
		(b) $0.65 (a modest increase)
		(c) $1.04 (doubling)
4.2	Mainstreaming of road pricing to increase revenue	(a) Only priced lanes/facilities are widely used
		(b) In addition to priced lanes/facilities, variable parking pricing and MBUF systems are used in some areas
4.3	User revenues raised per mile driven	(a) $34 per 1,000 VMT (essentially unchanged)
		(b) $45 per 1,000 VMT (30% increase)
4.4	Expenditures on roadways per mile driven	(a) $40 per 1,000 VMT (35% decrease)
		(b) $65 per 1,000 VMT (generally stable)
		(c) $80 per 1,000 VMT (30% increase)
4.5	Congestion	(a) Has increased only slightly
		(b) Has increased significantly

Table A.1. Influencing Areas, Descriptors, and Projections—Continued

	DESCRIPTOR	PROJECTION

Influencing area: Transportation funding and supply—continued

4.6	Quality and quantity of public transit	(a) Transit service (measured in revenue-miles) has increased by a total of 10%, and quality has increased
		(b) Transit service (measured in revenue-miles) has increased by a total of 10%, and quality has decreased[a]
		(c) Transit service (measured in revenue-miles) has increased by a total of 35%, and quality has increased[a]

Influencing area: Technology

5.1	Market penetration for broadband	(a) 75% of households use broadband technology
		(b) 90% of households use broadband technology
5.2	Telecommuting share	(a) 40% of workers telecommute
		(b) 15% of workers telecommute
5.3	Online shopping share of retail sales	(a) 30% of retail purchases (by number of transactions) are made online
		(b) 15% of retail purchases (by number of transactions) are made online
5.4	Development of data privacy regulations	(a) Strict regulation: Data privacy regulations do not allow the collection of in-vehicle data used, e.g., in MBUF systems or PAYD insurance
		(b) Weak regulation: Data privacy regulations allow the collection of in-vehicle data used, e.g., in MBUF systems or PAYD insurance
5.5	Adoption of telematic services	95% of all new vehicles in 2030
5.6	Market penetration of ADASs	(a) 90% of all new vehicles in 2030
		(b) 55% of all new vehicles in 2030
5.7	Market penetration of autonomous vehicles	(a) Essentially zero
		(b) Very low (no more than 5% share in total car fleet)

[a] The figures reported here for quality and amount of public transit do not match those projected in Tables 4.2 and 4.3 in Chapter Four because those projections were based on the methodology described earlier in this chapter, while the projections here were those produced at the expert workshops.

Step 3: Integrate into Scenario Frameworks

We performed two types of analysis to develop the input to the scenarios. First, we conducted a cross-impact analysis between the descriptors across all influencing areas. This identified the driving forces in the system (see Gausemeier, Fink, and Schlake, 1998, for a further description of this type of analysis). The impacts that the different descriptors have on each other were recorded in a cross-impact matrix (or influence matrix) using a scale from 0 (no impact) to 3 (strong impact). For example, the average household size has no direct impact on the oil price, so that relationship would be rated 0, while oil price has a strong direct impact on oil consumption, so it was rated 3. This exercise establishes the degree of interconnectedness of all descriptors.

The outcome of this analysis is the system diagram illustrated in Figure A.2. The higher the activity index of a descriptor, the more it influences other descriptors. For example, economic growth development affected a large number of other descriptors, so it is highly active. The higher the passivity index, the more a descriptor is driven by other descriptors. Oil consumption is affected by many other descriptors, so it is considered highly passive. Descriptors with both a high activity index and a high passivity index are strongly interconnected in the system, being driver and driven at the same time. This analysis was the basis for identifying some descriptors as key drivers.

The second type of analysis is based on consistency logic, which establishes consistency (or lack thereof) among projections across all descriptors. *Consistency* here means how well the projection of a particular row and column would "fit" and how realistic it would be for both of them to appear simultaneously. The matrix entry is a numerical value that represents the level of consistency, with 5 being strongly consistent and 1 being totally inconsistent. This was conducted in a workshop held in RAND's Santa Monica office in August 2012 that included both outside and RAND experts. We created a consistency matrix using all projections for each descriptor (see extract in Figure A.3). Workshop participants judged how a projection in a row is consistent or compatible with the projections in each column. A score of 1 indicates total inconsistency; for example, oil consumption of 20.6 million barrels per day (higher than the current rate) was judged totally inconsistent with adoption of a national GHG policy because we expect such a policy to reduce oil consumption. But the projection of a production level of 15 million barrels of oil per day with the consumption of 20.6 million barrels per day was rated a 5 for totally consistent because we expect higher levels of production and consumption to go hand in hand.

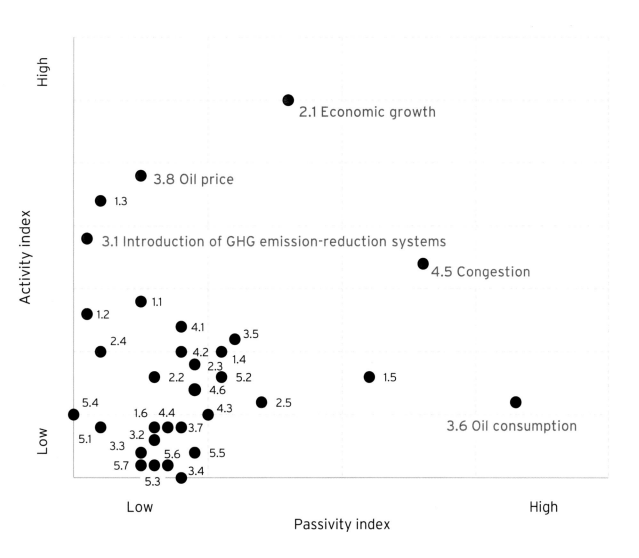

Figure A.2. System Dynamics as an Outcome of the Cross-Impact Analysis

NOTE: Numbers are cross-referenced to Table A.1. Key drivers are shown in orange.

		Introduction of GHG emission-reduction systems			Electricity power generation sources		Adoption of AFVs			Oil consumption		U.S. oil production	
		(a) National GHG policy was introduced by 2022	(b) No action had been taken in 2020, but policies were adopted by 2030	(c) No new GHG legislation has been adopted by 2030	(a) 20% share of nonhydro renewables	(b) 10% share of nonhydro renewables	(a) About 1% of all light-duty vehicles sold in 2030	(b) About 8% of all light-duty vehicles sold in 2030	(c) About 40% of all light-duty vehicles sold in 2030	(a) 16 million barrels per day	(b) 20.6 million barrels per day	(a) 15 million barrels per day	(b) 6 million barrels per day
Electricity power generation sources	(a) 20% share of nonhydro renewables	5	4	2									
	(b) 10% share of nonhydro renewables	1	2	4									
Adoption of AFVs	(a) About 1% of all light-duty vehicles sold in 2030	1	2	4	3	3							
	(b) About 8% of all light-duty vehicles sold in 2030	2	3	3	3	3							
	(c) About 40% of all light-duty vehicles sold in 2030	5	4	2	3	3							
Oil consumption	(a) 16 million barrels per day	5	4	2	3	3	2	3	5				
	(b) 20.6 million barrels per day	1	2	4	3	3	5	3	2				
U.S. oil production	(a) 15 million barrels per day	3	3	3	3	3	3	3	3	3	5		
	(b) 6 million barrels per day	4	3	3	3	3	3	3	3	4	3		
Oil price	(a) $90/barrel in 2030 (Brent crude in 2012 dollars)	3	3	3	3	3	5	4	3	1	4	2	5
	(b) $130/barrel in 2030 (Brent crude in 2012 dollars)	3	3	3	4	4	3	4	4	4	3	5	2
	(c) $190/barrel in 2030 (Brent crude in 2012 dollars)	3	3	3	5	4	2	5	5	5	2	5	2

Figure A.3. Extract from the Consistency Matrix, Including Projection Pairs

NOTE: Because this analysis is in one direction only (that is, how consistent a column projection is with a row projection), cells shaded in orange were not analyzed.

Rating scale

1 = Totally inconsistent

2 = Partially inconsistent

3 = Neutral or independent

4 = Consistent

5 = Strongly consistent

The consistency matrix was then fed into an online tool, the RAHS platform.[7] RAHS is a prototype of a web-based foresight platform that has been developed and funded by the Future Analysis Branch of the German Federal Ministry of Defence, to enhance internal and external cooperation and to strengthen the methodological fundamentals of its work. Instead of providing a single software solution only for scenario development, it supports foresight projects with a variety of alternative foresight methods within a Web 2.0 environment (Brockmann, 2012; Durst, Kolonko, and Durst, 2012). RAHS was designed based on a comprehensive scanning of internationally applied foresight methods and tools, including the Z_punkt Foresight-Toolbox, the Joint Research Centre (JRC) FOR-LEARN Online Foresight Guide, Foresight's Horizon Scanning Centre (HSC) toolkit, the European Union (EU) research project iKnow, and compilations of future research methodologies in the Millennium Project by Glenn and Gordon (2009) and Pillkahn (2007).

For this project, ifmo researchers led the use of RAHS to analyze millions of mathematically possible pairs of projections for the descriptors across all influencing areas and to eliminate the pairs deemed inconsistent in the consistency analysis that preceded this step. The exploratory scenario construction toolbox in RAHS isolated clusters made up of homogeneous groups of descriptors and projections based on the consistency analysis results. A complete linkage method was used to calculate distances between clusters. In complete linkage, the distance between two clusters was computed as the maximum distance between a pair of projections, one in one cluster and one in another. Developing the clusters was an agglomerative procedure. The clusters were initially single projection pairs (single-member clusters). Then pairs of projections that were closest according to the linkage criterion (that is, the consistency matrix value) were joined to form a new, larger cluster. At the last stage, a single cluster made up of all highly consistent projections was formed. (More details on the application of consistency logic and cluster analysis implemented in the RAHS platform can be found in Gausemeier, Fink, and Schlake, 1998.) From these, the research team used a two-step process to select two to develop further.

The first step was to identify projections that were deemed essential to scenario development—that is, projections that needed to appear in at least one scenario because of their importance. This identification was based on expert judgment at the workshop and was independent of the RAHS outputs. Each participant was asked to identify three specific projections of three descriptors that should be included in the final scenario set. The criterion for these critical projections is that they must have the potential to be highly relevant within a scenario. For example, one person might think that it is important to make sure to include a projection of $190-per-barrel oil in at least one scenario. Descriptors with only one projection were excluded because, by default, they would be included in both scenarios. Selected projections were written down on cards and pinned on a wall, and the experts voted on which were most important. We kept those projections that received three or more votes as critical projections; this narrowed the number to eight. The outcome of this process is summarized in Figure A.4.

---

[7] Although this platform is accessible online, it is only in German and requires a password to view.

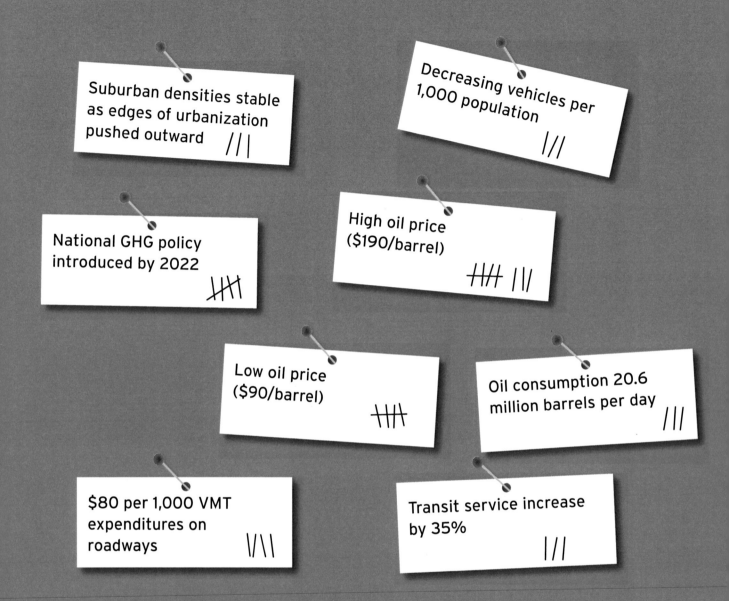

Figure A.4. Critical Projections for Selecting Scenario Frameworks

SOURCE: Institute for Mobility Research.

The second step was to identify two clusters of projections that together accounted for all eight projections. This was based on quantitative information derived from the RAHS tool.

The RAHS output (see Table A.2) enabled the experts to identify scenarios. For example, clusters 1 through 4 shared many common projections, while clusters 5 and 6 were relatively different. In particular, cluster 6 was an outlier with regard to some key projections. For example, in clusters 1 through 5, none of the combinations had an oil price of $90 per barrel, while all the combinations in cluster 6 had that price. Similarly, cluster 6 differed on related projections, such as cost to drive per mile (unchanged) and road pricing (only priced lanes are in widespread use). Cluster 6 thus became the Fueled and Freewheeling scenario.

Cluster 5 was selected because, in many respects, it was the furthest away from cluster 6, making for the greatest differentiation. For example, all the combinations in cluster 5 included the introduction of GHG legislation by 2022 (while, for clusters 1-4, most but not all included this projection). This pattern was similar for such projections as high electricity prices and high quantity and amount of transit supply. The one area in which clusters 5 and 6 were similar, in opposition to clusters 1-4, was economic growth; 5 and 6 included growth rates between 2 and 2.5 percent annually, while the others included slower growth rates. This was the topic of much debate because it would be more typical to include one higher-growth and one lower-growth scenario. However, both economic projections were for continued growth, so it was deemed less important to make this distinction. It was also of interest to produce a scenario that combined economic growth with GHG legislation.

Table A.2. Share of Projections Within Each Cluster (rounded)

Descriptor	Projection	Cluster					
		1	2	3	4	5	6
Economic growth	Slowed average annual growth between 1.5 and 2.0%	100	100	100	100	0	0
	Continued average annual growth between 2.0 and 2.5%	0	0	0	0	100	100
Income distribution	Income inequality continued to increase	100	100	100	100	100	100
Labor-force participation	Labor-force participation has remained stable, with men at about 70% and women at 60%	70	69	0	0	50	0
	Women's labor-force participation kept growing slightly while men's rate decreased slightly	30	31	100	100	50	100
Sector employment	Manufacturing continued to decline slightly but stabilized at 7-8%; service employment increased to 55-60%	100	100	100	100	100	100
Freight movement	Ton-miles increased annually by 0.6-0.8%	100	93	100	93	0	0
	Ton-miles increased annually by 0.8-1.0%	0	7	0	7	100	100
Total population	Moderate growth of 0.8% per annum	100	100	100	100	100	100
Share of population by race/ethnic group	Growing share of Hispanics and Asians; share of whites has continued to decline	100	100	100	100	100	100
Age structure	The share of the population over 65 has increased steadily	100	100	100	100	100	100
Population density	Urban/suburban densification with stable population shares	100	100	100	100	100	0
	Suburban densities stable as edges of urbanization pushed outward	0	0	0	0	0	100
Vehicles per 1,000 population	Increased	0	0	0	0	0	100
	Leveled off at 2009 rates	2	2	7	7	0	0
	Decreased	98	98	93	93	100	0
Average household size	Average household size has remained stable, but share of households with children decreased slightly	100	100	100	100	100	100
Introduction of GHG emission-reduction systems	National GHG policy was introduced by 2022	87	87	87	87	100	0
	No action had been taken in 2020, but policies were adopted by 2030	13	13	13	13	0	0
	No new GHG legislation has been adopted by 2030	0	0	0	0	0	100

Table A.2. Share of Projections Within Each Cluster (rounded)—Continued

Descriptor	Projection	Cluster					
		1	2	3	4	5	6
Electricity power generation sources	20% share of nonhydro renewables	100	100	100	100	100	0
	10% share of nonhydro renewables	0	0	0	0	0	100
EV-charging infrastructure	90-95% of EV charging is done at home; about 18,000 publicly available charging stations	6	7	7	7	0	100
	80-85% of EV charging is done at home; about 100,000 publicly available charging stations	94	93	93	93	100	0
Electricity prices	The average real price for all sectors is $0.13/kWh	6	7	7	7	0	100
	The average real price for all sectors is $0.18/kWh	94	93	93	93	100	0
Adoption of AFVs	About 1% of all light-duty vehicles sold in 2030	0	0	0	0	0	100
	About 8% of all light-duty vehicles sold in 2030	98	98	93	93	100	0
	About 30-45% of all light-duty vehicles sold in 2030	2	2	7	7	0	0
Oil consumption	16 million barrels/day	100	100	100	100	100	0
	20.6 million barrels/day	0	0	0	0	0	100
U.S. oil production	15 million barrels/day	13	13	13	13	100	100
	6 million barrels/day	87	87	87	87	0	0
Oil price	$90/barrel in 2030 (Brent crude in 2012 dollars)	0	0	0	0	0	100
	$130/barrel in 2030 (Brent crude in 2012 dollars)	6	7	7	7	0	0
	$190/barrel in 2030 (Brent crude in 2012 dollars)	94	93	93	93	100	0
Cost to drive per mile	$0.52 (unchanged)	0	0	0	0	0	100
	$0.65 (a modest increase)	6	7	7	7	0	0
	$1.05 (doubling)	94	93	93	93	100	0
Mainstreaming of road pricing to increase revenue	Only priced lanes/facilities are widely used	0	0	0	0	0	100
	In addition to priced lanes/facilities, variable parking pricing and MBUF systems are used in some areas	100	100	100	100	100	0

Table A.2. Share of Projections Within Each Cluster (rounded)—Continued

Descriptor	Projection	Cluster					
		1	2	3	4	5	6
User revenues raised per mile driven	$34 per 1,000 VMT (essentially unchanged)	0	0	0	0	0	100
	$45 per 1,000 VMT (30% increase)	100	100	100	100	100	0
Expenditures on roadways per mile driven	$40 per 1,000 VMT (35% decrease)	0	0	0	0	0	50
	$65 per 1,000 VMT (generally stable)	6	7	7	7	0	50
	$80 per 1,000 VMT (30% increase)	94	93	93	93	100	0
Congestion	Has increased only slightly	100	100	100	100	100	0
	Has increased significantly	0	0	0	0	0	100
Quality and amount of urban public transportation infrastructure	Transit service (measured in revenue-miles) has increased by a total of 10%, and quality increased	6	7	7	7	0	0
	Transit service (measured in revenue-miles) has increased by a total of 10%, and quality decreased	6	7	7	7	0	100
	Transit service (measured in revenue-miles) has increased by a total of 35%, and quality increased	87	87	87	87	100	0
Market penetration for broadband	75% of households use broadband technology	11	0	7	0	0	0
	90% of households use broadband technology	89	100	93	100	100	100
Telecommuting share	40% of workers telecommute	92	96	100	100	100	0
	15% of workers telecommute	9	4	0	0	0	100
Online shopping share of retail sales	30% of retail purchases (by number of transactions) are made online	0	100	1	100	100	100
	15% of retail purchases (by number of transactions) are made online	100	0	100	0	0	0
Development of data privacy regulations	Strict regulation: Data privacy regulations do not allow the collection of in-vehicle data used, e.g., in MBUF systems or PAYD insurance	6	7	7	7	0	50
	Weak regulation: Data privacy regulations allow the collection of in-vehicle data used, e.g., in MBUF systems or PAYD insurance	94	93	93	93	100	50

Table A.2. Share of Projections Within Each Cluster (rounded)—Continued

Descriptor	Projection	Cluster					
		1	2	3	4	5	6
Adoption of telematic services	95% of all new vehicles in 2030	100	100	100	100	100	100
Market penetration of ADASs	90% of all new vehicles in 2030	6	7	7	7	100	100
	55% of all new vehicles in 2030	94	93	93	93	0	0
Market penetration of autonomous vehicles	Essentially zero	36	36	100	100	0	0
	Very low (no more than 5% share in total car fleet)	64	64	0	0	100	100

SOURCE: RAHS analysis.

Step 4: Produce Scenario Narratives

Using the two selected scenario frameworks, the research team wrote a narrative for each scenario. The storylines were developed by interlinking active and passive descriptors. Key developments and interrelations were highlighted and interpreted. Thus, the scenarios describe not only the situation in 2030 but also how a situation developed step by step during that time frame. The scenarios represent a dynamic path, starting today and continuing to 2030. Using standard convention, all narratives were written from the vantage point of 2030.

Step 5: Draw Consequences for Future Mobility

To explore the consequences of the scenarios for future mobility, we derived numbers of PMT and mode shares for 2030 for different transport modes (vehicle, transit, domestic air, and intercity rail). We began by calculating the average annual per capita growth in each mode from 1990 to 2010 as a baseline, as shown in Table A.3.

Table A.3. Average Annual Per Capita Growth Rates in Passenger-Miles Traveled, 1990–2010 (%)

Transport Mode	Average Annual Growth
Vehicle, total	-0.21
Transit, total	0.14
Domestic air	1.37
Intercity rail	-0.79
Total, all modes	-0.05

SOURCE: BTS, undated (b), Table 1-40.
NOTE: *Highway*, the term used by BTS, refers to vehicular travel on all roadways, regardless of type.

Then, for each descriptor, we evaluated its impact on PMT going forward relative to its impact from 1990–2010. The impact was scored using a rating scale from -3 (strongly negative) to 3 (strongly positive). Zero means that a descriptor's future impact on PMT will be equal to its impact over the past 20 years. For example, the impact of population by race and ethnic group on highway PMT was evaluated as 0. From 1990 to 2010, the share of the population made up of Hispanics and Asians grew, and the share of whites declined. Because these trends are expected to continue through 2030, the future impact of these population trends should be similar to the past impact.

These PMT factors were subsequently multiplied by a relevance factor (from 0 to 3) to weight the impact of each descriptor. For example, the relevance of economic growth on highway PMT development was scored a 3 (strongly relevant), while the relevance of market penetration for broadband on highway PMT was rated 0 (not relevant). Summing up all weighted PMT factors and dividing by the sum of all relevance factors allowed us to calculate a normalized PMT factor for each transport mode.

Finally, this normalized PMT factor had to be retranslated into an average annual growth rate based on a translation matrix. We set the historical average annual PMT per capita growth rates as 0 in this matrix (see the average annual growth rates in Table A.4). Informed by past trends but independently of the developments summarized in the two scenarios, we defined -3 and 3 as the lowest and highest 20-year average annual growth rates that seemed plausible under very negative and very positive conditions, respectively. For example, we determined the highest imaginable growth rate for per capita vehicle travel to be 1.5 percent per year and a lowest to be -2.0 percent.

Although, in any single year, the total percentage increase or decrease might exceed these amounts, these figures were developed based on the plausible 20-year average. For example, even though total per capita highway travel fell by more than 15 percent from 2008 to 2009, it would be unrealistic to assume that it could fall by this much every year for 20 years. Using these anchors, we estimated future average annual growth rates for each transport mode and, thus, the calculation of absolute PMT numbers as presented in Chapter Four of this report.

Table A.4. Translation Matrix: Normalized Factor for Passenger-Miles Traveled and 20-Year Average Annual Growth Rates (%)

Transport Mode	Normalized PMT Factor						
	-3	-2	-1	0	1	2	3
Vehicle, total	-2.0	-1.40	-0.81	-0.21	0.36	0.93	1.5
Transit, total	-2.0	-1.29	-0.57	0.14	0.76	1.38	2.0
Domestic air	-1.0	-0.21	0.58	1.37	2.08	2.79	3.5
Intercity rail	-3.0	-2.26	-1.53	-0.79	0.14	1.07	2.0

Step 6: Create Wild-Card Scenarios

In scenario development, wild cards are highly unlikely but possible events that have a major impact on the future. They are disruptive and surprising, and they undermine the trends or developments presented in a scenario. During the workshops, the experts were asked to think about which wild cards would have a strong and sustained impact on future mobility in the United States. In the final workshop in Santa Monica, we selected two wild cards and discussed how they would break with developments described in the two scenarios. Informed by these discussions, we drafted the wild-card scenarios (Red Dusk: China Stumbles and The Autonomous-Vehicle Revolution) presented in Chapter Five of this report. They were also reviewed by RAND staff who were not involved with the workshops to ensure that, even though they are unlikely, the events described are plausible.

Appendix B
List of Experts

Table B.1 lists the outside experts who participated in each workshop, as well as their professional affiliations at the time the workshop took place. (The table does not include RAND and ifmo staff who were also present.) All five expert workshops were held in the RAND office in Arlington, Virginia. Table B.2 contains the participants in the cross-impact and consistency analysis workshop, held at RAND's Santa Monica office on August 6–7, 2012.

Table B.1. Expert Workshop Participants

Workshop Name and Date	Expert Name	Affiliation
Demographics, April 3, 2012	John Cromartie	Economic Research Service, U.S. Department of Agriculture
	Ryan Edwards	Queens College, City University of New York
	B. Lindsay Lowell	Georgetown University
	Joyce Manchester	Congressional Budget Office
	Nancy McGuckin	Independent consultant, travel behavior
	Jeffrey Passel	Pew Hispanic Center
Economics, April 5, 2012	Paul Bingham	CDM Smith
	Gregory Bischak	Community Development Financial Institutions Fund, U.S. Department of the Treasury
	Luca Flabbi	Georgetown University
	Andreas Kopp	World Bank
	Marika Santoro	Congressional Budget Office
	Sita Slavov	American Enterprise Institute
	Michael Toman	World Bank
	Jack Wells	U.S. Department of Transportation
Energy, June 8, 2012	Austin Brown	National Renewable Energy Laboratory, U.S. Department of Energy
	Carmine Difiglio	Office of Policy and International Affairs, U.S. Department of Energy
	Charles Ebinger	Brookings Institution
	Jim Kliesch	Union of Concerned Scientists
	Joshua Linn	Resources for the Future
	Michael Shelby	Office of Transportation and Air Quality, U.S. Environmental Protection Agency
	Jim Turnure	Energy Information Administration
	Jake Ward	Office of Energy Efficiency and Renewable Energy, U.S. Department of Energy
Technology, June 11, 2012	Steven Bayless	Intelligent Transportation Society of America
	Matthew Dorfman	D'Artagnan Consulting
	Frank Douma	University of Minnesota
	Philip Gott	IHS Global Insight
	Alain Kornhauser	Princeton University
	Greg Krueger	Science Applications International Corporation
	Ted Trepanier	INRIX

Table B.1. Expert Workshop Participants—Continued

Workshop Name and Date	Expert Name	Affiliation
Transportation funding and supply, June 13, 2012	Susan Binder	Cambridge Systematics
	John Fischer	Congressional Research Service (retired)
	Emil Frankel	Bipartisan Policy Center
	Art Guzzetti	American Public Transportation Association
	Philip Herr	Government Accountability Office
	Valerie Karplus	Massachusetts Institute of Technology
	Bruce Schaller	New York City Department of Transportation
	Mary Lynn Tischer	Federal Highway Administration

Although 27 experts provided input through the ExpertLens elicitation, the system is set up to provide anonymity to participants. We do know that we had participants from each influencing area: four in demographics, three in economics, nine in energy, six in transportation funding and supply, and five in technology.

Table B.2. Cross-Impact and Consistency Analysis Workshop Participants

Workshop Date	Expert Name	Affiliation
August 6-7, 2012	Johanna Zmud	RAND
	Liisa Ecola	RAND
	Peter Brownell	RAND
	Jan Osburg	RAND
	Thomas Light	RAND
	Costa Samaras	RAND
	Paul Sorensen	RAND
	Irene Feige	ifmo
	Peter Phleps	ifmo
	Nancy McGuckin	Independent consultant, travel behavior
	Paul Bingham	CDM Smith

References

Amer, Muhammad, Tugrul U. Daim, and Antonie Jetter, "A Review of Scenario Planning," *Futures*, Vol. 46, February 2013, pp. 23-40.

American Automobile Association, "Your Driving Costs: How Much Are You Really Paying to Drive? 2012 Edition," Heathrow, Fla., 2012. As of June 18, 2013:
http://newsroom.aaa.com/wp-content/uploads/2012/04/YourDrivingCosts2012.pdf

American Public Transportation Association, *Public Transportation Fact Book*, 63rd ed., Washington, D.C., September 2012. As of June 18, 2013:
http://www.apta.com/resources/statistics/Documents/FactBook/APTA_2012_Fact%20Book.pdf

Autor, David H., Lawrence F. Katz, and Alan B. Krueger, "Computing Inequality: Have Computers Changed the Labor Market?" *Quarterly Journal of Economics*, Vol. 113, No. 4, 1998, pp. 1169-1213.

Barboza, David, "Building Boom in China Stirs Fears of Debt Overload," *New York Times*, July 6, 2011. As of June 18, 2013:
http://www.nytimes.com/2011/07/07/business/global/building-binge-by-chinas-cities-threatens-countrys-economic-boom.html?pagewanted=all&_r=0

BEA—*See* Bureau of Economic Analysis.

Berman, Eli, John Bound, and Zvi Griliches, "Changes in the Demand for Skilled Labor Within U.S. Manufacturing Industries: Evidence from the Annual Survey of Manufacturers," *Quarterly Journal of Economics*, Vol. 109, No. 2, May 1994, pp. 367-397.

BLS—*See* Bureau of Labor Statistics

Borjas, George J., Richard B. Freeman, and Lawrence F. Katz, "On the Labor Market Effects of Immigration and Trade," in George J. Borjas and Richard B. Freeman, eds., *Immigration and the Work Force: Economic Consequences for the United States and Source Areas*, Chicago, Ill.: University of Chicago Press, 1992, pp. 213-244.

Borjas, George J., and Valerie A. Ramey, "Foreign Competition, Market Power, and Wage Inequality," *Quarterly Journal of Economics*, Vol. 110, No. 4, 1995, pp. 1075-1110.

Börjeson, Lena, Mattias Höjer, Karl-Henrik Dreborg, Tomas Ekvall, and Göran Finnveden, "Scenario Types and Techniques: Towards a User's Guide," *Futures*, Vol. 38, No. 7, September 2006, pp. 723-739.

Bound, John, and George Johnson, "Changes in the Structure of Wages in the 1980s: An Evaluation of Alternative Explanations," *American Economic Review*, Vol. 82, No. 3, June 1992, pp. 371-392.

Brockmann, Kathrin, "Futures Analysis Cooperation Tool in the German Armed Forces," *Foreknowledge*, No. 4, August 2012, pp. 6-7. As of June 18, 2013:
http://issuu.com/foreknowledge/docs/foreknowledge4/7?e=0

Brownell, Peter, Thomas Light, Paul Sorensen, Constantine Samaras, Nidhi Kalra, and Jan Osburg, *The Future of Mobility: Scenarios for the United States, Appendixes C-G*, Santa Monica, Calif.: RAND Corporation, RR-246/1-ifmo, 2013. As of October 2013:
http://www.rand.org/publications/research_reports/RR246.html

Bryant, Benjamin P., and Robert J. Lempert, "Thinking Inside the Box: A Participatory, Computer-Assisted Approach to Scenario Discovery," *Technological Forecasting and Social Change*, Vol. 77, No. 1, January 2010, pp. 34-49.

Bureau of Economic Analysis, "Personal Income Summary," Table SA1-3, undated. As of January 15, 2012:
http://www.bea.gov/iTable/iTable.cfm?ReqID=70&step=1&isuri=1&acrdn=4

——, *Measuring the Economy: A Primer on GDP and the National Income and Product Accounts*, Washington, D.C., September 2007. As of September 21, 2012:
http://www.bea.gov/national/pdf/nipa_primer.pdf

——, "Current-Dollar and 'Real' GDP," December 20, 2011, revision. Referenced March 2, 2012. As of June 18, 2013:
http://www.bea.gov/national/xls/gdplev.xls

Bureau of Labor Statistics, "Employees on Nonfarm Payrolls by Major Industry Sector, 1962 to Date," Table B-1, undated. Referenced September 21, 2012. As of June 18, 2013:
http://www.bls.gov/opub/ee/2012/ces/tableb1_201205.pdf

——, *Labor Force Characteristics by Race and Ethnicity, 2010*, Report 1032, August 2011. As of September 21, 2012:
http://www.bls.gov/cps/cpsrace2010.pdf

BTS—*See* Bureau of Transportation Statistics

Bureau of Transportation Statistics, "Freight Activity in the United States: 1993, 1997, 2002 and 2007," *National Transportation Statistics*, Table 1-58, undated (a). Referenced September 21, 2012. As of June 18, 2013:
http://www.rita.dot.gov/bts/sites/rita.dot.gov.bts/files/publications/national_transportation_statistics/html/table_01_58.html

——, "Number of U.S. Aircraft, Vehicles, Vessels, and Other Conveyances," *National Transportation Statistics*, Table 1-11, undated (b). Referenced May 27, 2012. As of June 18, 2013: http://www.rita.dot.gov/bts/sites/rita.dot.gov.bts/files/publications/national_transportation_statistics/html/table_01_11.html

Crane, Keith, Andreas Goldthau, Michael Toman, Thomas Light, Stuart E. Johnson, Alireza Nader, Angel Rabasa, and Harun Dogo, *Imported Oil and U.S. National Security*, Santa Monica, Calif.: RAND Corporation, MG-838-USCC, 2009. As of July 11, 2013: http://www.rand.org/pubs/monographs/MG838.html

Davis, Stacy C., Susan W. Diegel, and Robert G. Boundy, *Transportation Energy Data Book, Edition 30*, Oak Ridge, Tenn.: Center for Transportation Analysis, Oak Ridge National Laboratory, 2011.

——, *Transportation Energy Data Book: Edition 31*, Oak Ridge, Tenn.: Oak Ridge National Laboratory, ORNL-6987, July 2012. As of June 18, 2013: http://cta.ornl.gov/data/index.shtml

DiNardo, John, Nicole M. Fortin, and Thomas Lemieux, "Labor Market Institutions and the Distribution of Wages, 1973-1992: A Semi-Parametric Approach," *Econometrica*, Vol. 64, No. 5, September 1996, pp. 1001-1044.

Durst, Carolin, Thomas Kolonko, and Michael Durst, "Kooperationsdilemma in der Zukunftsforschung Ein IT-basierter Lösungsansatz der Bundeswehr" [Cooperation dilemma in future analysis: A web-based approach by the German Armed Forces], [in German], *Proceedings Multikonferenz Wirtschaftsinformatik*, Braunschweig, February 29, 2012.

Economics and Statistics Administration and National Telecommunications and Information Administration, *Exploring the Digital Nation: Computer and Internet Use at Home*, Washington, D.C.: U.S. Department of Commerce, November 9, 2011. As of June 18, 2013: http://www.ntia.doc.gov/report/2011/exploring-digital-nation-computer-and-internet-use-home

EIA—*See* U.S. Energy Information Administration.

Executive Committee, Transportation Research Board, *Critical Issues in Transportation, 2009 Update*, Washington, D.C.: National Academies Press, c. 2009. As of June 18, 2013: http://www.trb.org/Publications/PubsCriticalIssuesinTransportation.aspx

Federal Highway Administration, *2002 Status of the Nation's Highways, Bridges, and Transit: Conditions and Performance— Report to Congress*, Washington, D.C., c. 2003. As of June 18, 2013: http://www.fhwa.dot.gov/policy/2002cpr/index.htm

——, *2010 Status of the Nation's Highways, Bridges, and Transit: Conditions and Performance*, Washington, D.C., 2010. As of June 28, 2013: https://www.fhwa.dot.gov/policy/2010cpr/

Feenstra, Robert C., and Gordon H. Hanson, "Globalization, Outsourcing and Wage Inequality," *American Economic Review*, Vol. 86, No. 2, May 1996, pp. 240-245.

FHWA—*See* Federal Highway Administration.

Flyvbjerg, Bent, "Survival of the Unfittest: Why the Worst Infrastructure Gets Built—and What We Can Do About It," *Oxford Review of Economic Policy*, Vol. 25, No. 3, 2009, pp. 344-367.

Freeman, Richard B., "Labor Market Institutions and Earnings Inequality," *New England Economic Review*, May-June 1996, pp. 157-168.

Gausemeier, Juergen, Alexander Fink, and Oliver Schlake, "Scenario Management: An Approach to Develop Future Potentials," *Technological Forecasting and Social Change*, Vol. 59, No. 2, October 1998, pp. 111-130.

Gerst, M. D., P. Wang, and M. E. Borsuk, "Discovering Plausible Energy and Economic Futures Under Global Change Using Multidimensional Scenario Discovery," *Environmental Modelling and Software*, Vol. 44, June 2013, pp. 76-86.

Glenn, Jerome C., and Theodore J. Gordon, *Futures Research Methodology, Version 3.0*, Washington, D.C.: Millennium Project, 2009. As of April 4, 2013: http://www.millennium-project.org/millennium/FRM-V3.html

Gordon, Theodore J., and H. Hayward, "Initial Experiments with the Cross Impact Matrix Method of Forecasting," *Futures*, Vol. 1, No. 2, December 1968, pp. 100-116.

Haub, Carl, "The U.S. Recession and the Birth Rate," Washington, D.C.: Population Reference Bureau, July 2009. As of February 20, 2012: http://www.prb.org/Articles/2009/usrecessionandbirthrate.aspx

Hobbs, Frank, and Nicole Stoops, "Demographic Trends in the 20th Century," Washington, D.C.: U.S. Census Bureau, Census 2000 Special Report 4, November 2002. As of June 18, 2013: http://www.census.gov/prod/2002pubs/censr-4.pdf

Horrigan, John, *Online Shopping*, Washington, D.C.: Pew Internet and American Life Project, February 13, 2008. As of May 9, 2012: http://www.pewinternet.org/Reports/2008/Online-Shopping.aspx

Howden, Lindsay M., and Julie A. Meyer, "Age and Sex Composition: 2010," Washington, D.C.: U.S. Census Bureau, 2010 Census Brief 3, May 2011. As of June 18, 2013: http://www.census.gov/prod/cen2010/briefs/c2010br-03.pdf

Internal Revenue Service, *Travel, Entertainment, Gift, and Car Expenses*, Washington, D.C., various years. As of June 18, 2013: http://www.irs.gov/uac/ Publication-463,-Travel,-Entertainment,-Gift,-and-Car-Expenses-1

Kahn, Herman, and Thomas Pepper, *The Japanese Challenge: The Success and Failure of Economic Success*, New York: Crowell, 1979.

Kahn, Herman, and Anthony J. Wiener, "The Next Thirty-Three Years: A Framework for Speculation," *Daedalus*, Vol. 96, No. 3, Summer 1967, pp. 705-732.

Kane, Julius, "A Primer for a New Cross-Impact Language: KSIM," *Technological Forecasting and Social Change*, Vol. 4, No. 2, 1972, pp. 129-142.

Katz, Lawrence F., and Kevin M. Murphy, "Changes in Relative Wages, 1963-1987: Supply and Demand Factors," *Quarterly Journal of Economics*, Vol. 107, No. 1, February 1992, pp. 35-78.

Lee, David S., "Wage Inequality in the United States During the 1980s: Rising Dispersion or Falling Minimum Wage?" *Quarterly Journal of Economics*, Vol. 114, No. 3, 1999, pp. 977-1023.

Lempert, Robert J., Steven W. Popper, and Steven C. Bankes, *Shaping the Next One Hundred Years: New Methods for Quantitative, Long-Term Policy Analysis*, Santa Monica, Calif.: RAND Corporation, MR-1626-RPC, 2003. As of June 18, 2013: http://www.rand.org/pubs/monograph_reports/MR1626.html

Mietzner, Dana, and Guido Reger, "Advantages and Disadvantages of Scenario Approaches for Strategic Foresight," *International Journal of Technology Intelligence and Planning*, Vol. 1, No. 2, 2005, pp. 220-239.

Miniño, Arialdi M., Sherry L. Murphy, Jiaquan Xu, and Kenneth D. Kochanek, "Deaths: Final Data for 2008," *National Vital Statistics Reports*, Vol. 59, No. 10, December 7, 2011. As of June 18, 2013: http://www.cdc.gov/nchs/data/nvsr/nvsr59/nvsr59_10.pdf

Moreno, Robert B., and Peter Zalzal, "Greenhouse Gas Dissonance: The History of EPA's Regulations and the Incongruity of Recent Legal Challenges," *Journal of Environmental Law*, Vol. 30, No. 121, 2012, pp. 121-156.

Murphy, Kevin M., and Finis Welch, "The Structure of Wages," *Quarterly Journal of Economics*, Vol. 107, No. 1, February 1992, pp. 285-326.

Murphy, Sherry L., Jiaquan Xu, and Kenneth D. Kochanek, "Deaths: Preliminary Data for 2010," *National Vital Statistics Reports*, Vol. 60, No. 4, January 11, 2012. As of February 8, 2013: http://www.cdc.gov/nchs/data/nvsr/nvsr60/nvsr60_04.pdf

National Highway Traffic Safety Administration, "Obama Administration Finalizes Historic 54.5 mpg Fuel Efficiency Standards," press release, Washington, D.C., August 28, 2012. As of September 19, 2012: http://www.nhtsa.gov/About+NHTSA/Press+Releases/2012/ Obama+Administration+Finalizes+Historic+54.5+mpg+Fuel+ Efficiency+Standards

Nguyen, David H., and Gillian R. Hayes, "Information Privacy in Institutional and End-User Tracking and Recording Technologies," *Journal of Personal and Ubiquitous Computing*, Vol. 14, No. 1, January 2010, pp. 53-72.

Office of Energy Efficiency and Renewable Energy, "Alternative Fueling Station Total Counts by State and Fuel Type," updated April 30, 2012. Referenced May 27, 2012. As of June 18, 2013: http://www.afdc.energy.gov/afdc/fuels/stations_counts.html

Pillkahn, Ulf, *Trends und Szenarien als Werkzeuge zur Strategieentwicklung* [Trends and scenarios as instrument for strategy development], [in German], Erlangen: Publicis Publishing, 2007.

Rozenberg, Julie, Céline Guivarch, Robert Lempert, and Stéphane Hallegatte, *Building SSPs for Climate Policy Analysis: A Scenario Elicitation Methodology to Map the Space of Possible Future Challenges to Mitigation and Adaptation*, Fondazione Eni Enrico Mattei Working Paper 52.2012, 2012. As of June 18, 2013: http://www.feem.it/ getpage.aspx?id=4951&sez=Publications&padre=73

113

Ruggles, Steven, J. Trent Alexander, Katie Genadek, Ronald Goeken, Matthew B. Schroeder, and Matthew Sobek, *Integrated Public Use Microdata Series: Version 5.0*, Minneapolis, Minn.: University of Minnesota, 2010. As of June 18, 2013: https://usa.ipums.org/usa/index.shtml

Schweizer, Vanessa Jine, and Elmar Kriegler, "Improving Environmental Change Research with Systematic Techniques for Qualitative Scenarios," *Environmental Research Letters*, Vol. 7, No. 4, 2012.

Shrank, David, and Tim Lomax, *The 2011 Urban Mobility Report*, College Station, Texas: Texas Transportation Institute, c. 2012.

Toh, Han Shih, "Stimulus Scheme a Massive Debt Burden on Provinces," *South China Morning Post*, October 18, 2012.

U.S. Census Bureau, "Average Household Size of Occupied Housing Units by Tenure," *2010 American Community Survey 1-Year Estimates*, Table B25010, c. 2010a. As of June 18, 2013: http://factfinder2.census.gov/faces/tableservices/jsf/pages/productview.xhtml?pid=ACS_10_1YR_B25010&prodType=table

——, "Household Size by Vehicles Available," *2010 American Community Survey 1-Year Estimates*, Table B08201, c. 2010b.

——, "Population, Housing Units Area, and Density: 2010 – United States - Metropolitan and Micropolitan Statistical Area Population by Size Class," American Fact Finder, Summary File 1, GCT-PH1, c. 2010c.

——, "Selected Measures of Household Income Dispersion: 1967 to 2010," Table A-3, c. 2010d. As of September 21, 2012: http://www.census.gov/hhes/www/income/data/historical/inequality/IE-1.pdf

——, "Civilian Labor Force and Participation Rates with Projections: 1980 to 2018," *Statistical Abstract of the United States 2012*, Table 587, January 2010e. As of September 21, 2012: http://www.census.gov/compendia/statab/cats/labor_force_employment_earnings.html

——, "Money Income of Families: Number and Distribution by Race and Hispanic Origin–2009," *Statistical Abstract of the United States: 2012*, Table 695, c. 2011a. As of June 19, 2013: http://www.census.gov/compendia/statab/2012/tables/12s0695.pdf

——, "Regions of People by Median Income and Sex: 1953 to 2010," Table P-5, c. 2011b. As of September 21, 2012: http://www.census.gov/hhes/www/income/data/historical/people/

——, *Statistical Abstract of the United States: 2012*, c. 2011c. As of June 4, 2012: http://www.census.gov/compendia/statab/

U.S. Energy Information Administration, *Annual Energy Outlook 2010 with Projections to 2035*, Washington, D.C., DOE/EIA-0383(2010), April 2010. As of July 12, 2013: http://www.eia.gov/oiaf/aeo/pdf/0383(2010).pdf

——, *Annual Energy Outlook 2011 with Projections to 2035*, Washington, D.C., DOE/EIA-0383(2011), April 26, 2011a. As of June 18, 2013: http://www.eia.gov/forecasts/archive/aeo11/index.cfm

——, "Short-Term Energy Outlook: Real Prices Viewer," December 6, 2011b. Referenced December 19, 2011. As of June 18, 2013: http://www.eia.gov/forecasts/steo/realprices/

——, *Annual Electric Power Industry Report*, Washington, D.C., Form EIA-861 Data Files, released September 20, 2012. Referenced September 20, 2012. As of June 18, 2013: http://www.eia.gov/electricity/data/eia861/index.html

Van der Hiejden, Kees, *Scenarios: The Art of Strategic Conversation*, New York: Wiley and Sons, 2005.

Wood, Adrian, "How Trade Hurt Unskilled Workers," *Journal of Economic Perspectives*, Vol. 9, No. 3, Summer 1995, pp. 57-80.

World Bank, "World Development Indicators," December 2011. Referenced June 4, 2012. As of June 18, 2013: http://data.worldbank.org/data-catalog/world-development-indicators/wdi-2011

Figures and Tables

Abbreviations

AAA	American Automobile Association
ACS	American Community Survey
ADAS	advanced driver-assistance system
AFV	alternatively fueled vehicle
APTA	American Public Transportation Association
BEA	Bureau of Economic Analysis
BEV	battery electric vehicle
BLS	Bureau of Labor Statistics
BRIC	Brazil, Russia, India, and China
BTS	Bureau of Transportation Statistics
BYOD	bring your own device
CAFE	Corporate Average Fuel Economy
CO_2	carbon dioxide
CPI	Consumer Price Index
E85	a mixture of gasoline and up to 85 percent ethanol
EERE	Office of Energy Efficiency and Renewable Energy
EIA	U.S. Energy Information Administration
ESA	Economics and Statistics Administration
ETS	emission trading scheme
EU	European Union
EV	electric vehicle
FCV	fuel-cell vehicle
FHWA	Federal Highway Administration
GDP	gross domestic product
GHG	greenhouse gas
GPS	global positioning system
HEV	hybrid electric vehicle
HOT	high-occupancy toll
HSC	Horizon Scanning Centre
ifmo	Institute for Mobility Research
IRS	Internal Revenue Service
JRC	Joint Research Centre
kWh	kilowatt-hour
MBUF	mileage-based user fee
mpg	mile per gallon
NHTSA	National Highway Traffic Safety Administration
NTIA	National Telecommunications and Information Administration
PAYD	pay as you drive
PBC	People's Bank of China
PHEV	plug-in hybrid electric vehicle
PMT	passenger-mile traveled
RAHS	Risk Assessment and Horizon Scanning
RFS	Renewable Fuel Standard
SIPP	Survey of Income and Program Participation
SOE	state-owned enterprise
TRB	Transportation Research Board
tWh	terawatt-hour
UNFCCC	United Nations Framework Convention on Climate Change
VMT	vehicle-mile traveled